Student Hints and Solutions Manual

to accompany

Mathematics

for

Elementary Teachers

A CONTEMPORARY APPROACH

Third Edition

Gary L. Musser and William F. Burger
Oregon State University, Corvallis, OR

Prepared by
Lynn E. Trimpe
Linn-Benton Community College, Albany, OR
Roger J. Maurer
Linn-Benton Community College, Albany, OR
Vikki R. Maurer

Macmillan College Publishing Company
New York
Maxwell Macmillan Canada
Toronto
Maxwell Macmillan International
New York   Oxford   Singapore   Sidney

Macmillan College Publishing Company
866 Third Avenue, New York, New York 10022

Macmillan College Publishing Company is part of the
Maxwell Communications Group of Companies.

Maxwell Macmillan Canada, Inc.
1200 Eglinton Avenue East, Suite 200
Don Mills, Ontario M3C 3N1

ISBN   0-02-421514-7

Printing: 1 2 3 4 5 6 7 8   Year: 4 5 6 7 8 9 0 1 2 3

PREFACE

This manual contains hints, additional hints, and complete solutions for all of the Part A problems in the text, *Mathematics for Elementary Teachers*, by Gary L. Musser and William F. Burger. It is designed to help you improve your problem-solving ability by providing hints to get you started on each of the problems, as well as complete solutions that model one correct solution for each problem.

How to Use this *Student Hints and Solutions Manual*

Your textbook contains some challenging problems. You will probably be able to solve many of them without referring to this resource, but there will be some that may "stump" you at first. To make the best use of this resource, you should make your best effort to solve a problem before you turn to the solutions manual. The only way to become a better problem-solver is by solving problems yourself. Therefore, you should first try solving each problem by referring to the suggestions and examples given in class and in your text. If, after a reasonable period of time, you are unsuccessful in your efforts to solve a problem, we suggest following these steps:

1) Go on to another problem or take a break from your studying. Come back to the problem later and give it another try.

2) Read the hint for that problem in Part 1 of this manual, and try the problem again.

3) Read the additional hint for the problem (if there is one), and try again.

4) Read *part* of the solution to the problem in Part 2 of this manual, and see if you can complete the solution yourself.

5) Read the rest of the solution, and see if you can determine where your difficulty lay.

NOTE: Even if you have solved a problem correctly, you may find this resource useful, as it may show you another correct solution to the problem or a completely different approach to the problem.

Acknowledgments

We wish to thank the following people for their invaluable assistance in the production of the *Student Hints and Solutions Manual*: Gary Musser for his technical expertise and moral support, Rosemary Troxel for proofreading, Scott Erwert and Aimee Schweizer for graphics, and Zak Zimmer and June Merzenich for word processing.

Lynn E. Trimpe
Roger J. Maurer
Vikki R. Maurer

CONTENTS

PART 1 - HINTS FOR PART A PROBLEMS

from

Mathematics for Elementary Teachers

A CONTEMPORARY APPROACH

by Gary L. Musser and William F. Burger

prepared by Lynn E. Trimpe

HINTS - PART A PROBLEMS

Chapter 1
Section 1.1

1. There are two different-sized triangles possible.

2. Use a variable to represent the width of the tennis court.

3. The square of what one-digit number is also a perfect cube?

4. Any even number can be represented as $2n$, for some integer n. How could you represent an odd number?

5. Try the guess and test strategy. *

6. Use variables to represent the numbers in the circles. Consider whether the sums in the squares are even or odd.

7. Draw pictures of the possible paths. Be systematic in the routes you try.

8. Fill the smaller, 3-liter pail first.

9. Draw a picture. *

10. Draw a picture and try the guess and test strategy.

11. Use systematic guess and test strategy. Try starting with possibilities for the corners first.

12. Notice that the circle in the upper left has no arrows leaving it. What can you conclude from that fact?

13. Try the guess and test strategy.

14. Try using a variable to represent each of the original numbers. You can then write two equations. *

15. Try the guess and test strategy, together with a picture.

16. (a) You want larger digits in high place-value positions.
    (b) You want to find two three-digit numbers that are as close to the same value as possible.

17. Use guess and test. *

18. Look for two (or more) squares whose sides together make up the side of another square. *

19. Use two variables. *

20. Use systematic guess and test. Since you are asked the man's age in 1949, you know he must have been born prior to 1949. *

21. Try systematic guess and test. You might first consider the possible values for the letter P. *

22. Consider the total amount of lost wages due to the strike and the number of hours Kathy works during one year. *

23. Draw a grid and use guess and test. Record the number of moves you make for each try.

24. The six-minute period may not start until one or both timers have gone off at least once. *

25. Try using variables or drawing a picture.

An additional hint for this problem is given in the next section.

26. Draw a picture and use guess and test.

27. Use a variable to represent the original number and perform the given operations.

28. Use the guess and test strategy. *

29. Think about who must ride the elevator first. Drawing a picture might help.

30. You might make a list of the perfect cubes first.

Section 1.2

1. Compare the number of terms in the sum to the answer.

2. Compare successive terms in each sequence and look for a pattern. *

3. Compare successive figures and look for patterns in each component of the drawing.

4. Consider the number of digits in each line of the sequence.

5. What arithmetic operations will make true statements with the numbers in each row?

6. (f) It might be helpful to add a third column to the table you made for part (a). In the new column list the number that is twice the number of dots in the figure. How is this number related to the triangular number?

7. A three-column table may be helpful here. Let the columns be: n, number of dots in the nth

triangular number, and number of dots in the nth square number.

8. Imagine section A of the newspaper opened flat. What numbers would appear on the full pages directly above and directly below the page described?

9. Try rewriting the sum inside the parentheses in each case. *

10. Try comparing two coins at a time, one per pan. What are the possible outcomes and what conclusions can you draw?

11. What do all of the amounts paid have in common; i. e., what kind of numbers are they?

12. How does each perimeter compare to the number of triangles?

13. Make a table comparing dimensions and perimeters.

14. Use systematic guess and test. Think about what numbers are multiples of 5, 3, or 6.

15. You might start by counting the zeros that will precede nonzero digits.

16. Look for a pattern in the columns of numbers. Where are the even numbers located? Where are the odd numbers located? *

17. There will be three different sizes of triangles.

18. Write out the dimensions of each pile of cubes.

19. Compare the factors in the product with the terms in the sum and with

---

**An additional hint for this problem is given in the next section.*

the terms in the Fibonacci sequence.

20. How does the sum in each example relate to the Fibonacci sequence?

21. Look for a pattern in the sequence of sums.

22. Be sure to draw some circles on the edge of the triangle as well as some in the center of the triangle.

23. Use the formula $D = R \times T$ to calculate the total time for a round trip in each case.

24. (d) Try writing the number of squares at each step as a sum, rather than as a single term. That is, write $1 + 4 + 8$ rather than 13 at step 3. *

25. Make a list of the positive integers. Determine which of those integers can be written as a sum of 4's and 9's and which ones cannot. *

26. Count the small points on each larger point of the star.

Chapter 2
Section 2.1

24. Draw various Venn diagrams.

25. Draw a diagram showing the possible pairings. *

26. Make up sets with the appropriate number of elements and list all possible subsets in each case. *

27. What must be true in order for two sets to be equal? To be equivalent?

28. Draw a Venn diagram or make up several examples of sets D and E.

29. Write out a set of eight skirts and a set of seven blouses. Then find the Cartesian product of the two sets you have written.

30. Try a simpler problem. Start with fewer entries first; for example, suppose there were only five entries.

31. Draw a Venn diagram of these three sets: TV watchers, newspaper readers, and radio listeners. Indicate the number of persons in each region of the diagram. *

32. Draw a Venn diagram of the three sets A, B, and C and indicate the number of persons in each region.

33. You must find a systematic way to pair up each point on $\overline{AC}$ and $\overline{BC}$ with a unique point on $\overline{AB}$. *

34. Draw a picture. You must find a way to connect each point on the chord to a unique point on the arc of the circle.

Section 2.2

15. When is the additive principle applied and when is the subtractive principle applied in the Roman numeration system? Does position matter in the Egyptian system?

16. (a) Write out the corresponding Egyptian numerals and count the symbols used.
    (b) Look for a pattern in your answers to (i) - (iv) in part (a).

An additional hint for this problem is given in the next section.

17. When would a 1993 model car first be advertised?

Section 2.3

19. Look at a simpler problem first. For example, how many digits would be used for a book of 25 pages?

20. Use the expanded form of the numeral. *

21. Represent the month, day, and year of birth as two-digit numbers in expanded notation. For example, let the month be $10a + b$.

22. (a)-(d) Write numerals on both sides of the equation in expanded form. *

23. (a)-(d) Make a list of numerals in the given base and examine the ones digits.

24. Solve a simpler problem. Consider two 3-digit numbers, one in base two and one in base ten. Which could a friend guess, digit-by-digit, in the fewest guesses?

Section 2.4

11. (c) If degrees Celsius equals degrees Fahrenheit, then $g(m) = m$, and $f(n) = n$. Solve one of these equations.

12. Look at the difference between successive terms.

13. (c) You want $C(x) > 1000$. Consider only whole numbers of months.

14. (a) Use the formula for the terms of a geometric series and the

two given terms to write two equations in two unknowns.

15. (c) Use the formula for the general term of a sequence of this type to write your function $T(n)$.

16. (b) Is there a common ratio or difference between successive terms?
    (c) Consider the number of triangles formed in each step and the number of toothpicks in each triangle. *

17. Use the formula for the general term of an arithmetic or geometric sequence to write the function $A(n)$.

18. Use the formula for the general term of an arithmetic or geometric sequence to write the function $A(n)$.

19. The flight of the clown ends when she lands on the ground, that is, when $h(t) = 0$.

Chapter 3
Section 3.1

16. Remember that if a set is closed under addition, the sum of any two elements of the set (not necessarily different elements) must also be an element of the set.

17. Since only three signs can be used, at least some of the numerals must have more than one digit.

18. Remember that the sums of all four rows are the same and that the numbers 10 - 25 will fill the grid. Use these facts to determine the sum in any row of the magic square. *

An additional hint for this problem is given in the next section.

19. What is "magic" about a magic square is the common sums. Use the numbers given in the hexagon to calculate a few sums. Do you see any patterns?

20. (b) Notice that if no "carries" are involved, a palindrome results after one addition. Try a 2-digit number that will involve "carries." *

21. Let n = youngest daughter's age. What algebraic expression would represent the next youngest daughter's age?

22. If you add the number of elements in A to the number of elements in B, which elements were counted twice?

23. Can you find examples of whole numbers a, b, and c for which $a + c = b + c$ but $a \neq b$?

Section 3.2

17. Remember that multiplication can be thought of as repeated addition.

18. (a) To find the sum of the numbers in a row, first find the sum of all the numbers in the magic square.
    (b) Choose a base and use digits 1 - 9 as exponents. *

19. Look for a pattern in the number and type of whole numbers added together.

20. Use systematic guess and test. Start by considering which products are possible.

21. Let n = original number. Write an algebraic expression representing

the operations performed on the number.

22. Use direct reasoning. A table or a list is helpful for keeping track of possibilities that you eliminate.

23. Use variables or use the guess and test strategy. *

24. The guess and test strategy works well here. *

25. Make a list comparing the number of days elapsed and the number of creatures on Earth.

26. Use direct reasoning. See solution to initial problem.

27. Can you find examples of whole numbers a, b, and c for which $ac = bc$ but $a \neq b$?

Section 3.3

11. (a) Write 6 as a product of factors and use properties of exponents.
    (b) Write 9 as a product of factors and use properties of exponents.
    (c) Write 12 as a product of factors and use properties of exponents.

12. Make a table listing the number of five-year periods since 1980 in one column and the price of a candy bar at that time in another column of the table. *

13. Remember that if $n(A) < n(B)$, then A matches a proper subset of B. *

14. What pattern do you notice in the sum of two cubes on the right-hand side? What patterns do you

An additional hint for this problem is given in the next section.

observe in the terms that appear on the left-hand side? *

15. Each size of pizza could have 0, 1, 2, 3, or all 4 toppings on it. *

16. Use systematic guess and test. A list of the perfect squares would be helpful here.

17. Consider all possible products in each case and compare answers to perfect squares.

18. Write the problem in general using expanded form. For example, $10a + b$ would represent the 2-digit number. *

19. (a) If $a < b$, then $a + n = b$ for some nonzero whole number n. *
    (b) Rewrite the Property of Less Than and Addition for the operation of subtraction by replacing each addition with subtraction.

Chapter 4
Section 4.1

28. Use the guess and test strategy. *

29. You want a low estimate and a high estimate for the differences.

30. Rewrite 13,333,333 as a sum and use distributive property to find its square.

31. Try writing 99 as $100 - 1$.

32. Compare the tens digit in the number to be squared with the first two digits in the answer. Do you see a relationship? *

33. You might first calculate the number of minutes required for sound to reach the moon.

34. Look at the last digit (ones digit) of each of the factors.

35. Rewrite each factor in the product by adding to or subtracting from the same number.

36. Rewrite each number on the right as a sum and then multiply using the distributive property.

37. Rewrite the larger number as a sum and then apply the distributive property.

38. Be careful with the order of operations.

39. Compare the digits in the products to the digits in each factor. *

40. To use the "round a 5 up" method, what digit of the number do you look at? *

Section 4.2

22. Work the problem correctly yourself, and compare your steps with the "student" versions.

23. Try systematic guess and test. Many answers are possible. *

24. Where must the largest digits in the set be placed in order to yield the largest sum? To yield the smallest sum?

25. Use systematic guess and test. Start with two-digit numbers whose sum is close to 100.

---

* An additional hint for this problem is given in the next section.

26. Use systematic guess and test. Notice that the digits remaining in a column (plus any "carries") must add up to 1, 11, or 21.

27. (d) Try other pairs of numbers such as 21 and 52, or 22 and 51. Look at sums and differences again. Consider also products of such pairs of numbers. Do you see a pattern?

28. Compare the number of rows and columns in the figure to the terms in the sum.

29. Use systematic guess and test. Notice that not all digits 0 - 9 can be used and that the possibilities for R are limited. *

30. Work each problem yourself, and compare your steps to the answer obtained by each student. Look for a pattern in each student's answers.

31. Notice that each digit in one factor is multiplied by each digit in the other factor, and each of the resulting products occupies a particular place-value position.

Section 4.3

9. Look at the digits used and the sum of the digits in the units places. *

10. Let x = the amount of money Steve has.

11. What digits can appear in the units place of a perfect square?

12. Make a list of perfect squares larger than 100. When the difference

between one of these squares and 100 is added to 164, is the result a perfect square?

Chapter 5
Section 5.1

14. Let $n = a \times 10^2 + b \times 10 + c$. You want to show that 5 divides n if and only if 5 divides c. *

15. Write out or imagine what each factorial looks like. For example, $20! = 20 \times 19 \times 18 \times \ldots \times 3 \times 2 \times 1$. You do *not* need to multiply the factors together to get a numerical answer. *

16. (a)-(b) See the hint for problem 15.
    (c) How do the numbers 8 and 7 differ?

17. What are the prime numbers less than 30? Check to see which of them is a factor of one of the given numbers.

18. Continue to substitute whole numbers for n. You will not need to check any counting numbers greater than 17 since $p(17) = 17^2 + 17 + 17$ is not prime.

19. (a) List your results in a table for reference when you do part (b). Use divisibility tests to check for primes.

20. One of the tests for divisibility will give you a factor of a group of these numbers. *

21. Consider the fact that, except for 2, all the prime numbers are odd. So the sum of two primes, both different from 2, would be even.

* An additional hint for this problem is given in the next section.

22. A list of the first 10 or 12 perfect squares might be useful here.

23. Keep in mind that of two consecutive counting numbers, one must be even and one must be odd.

24. Use the tests for divisibility to check for prime factors.

25. (b) Think about the fact that each odd number can be obtained by adding an odd number to an even number.

26. To save time in testing pairs of numbers, notice that the same two primes may work for several pairs. For example, the primes 11 and 13 lie between 7 and 14 and also between 8 and 16.

27. Use the prime factorization of 1,234,567,890 to determine all of its possible factors.

28. Examine the kind of factors each product in the set has. Do you see any common factors?

29. Look at the prime factorizations of the numbers 2 through 10. What different factors must a multiple of these numbers have?

30. Look at the prime factorizations of the numbers 2, 4, 5, 6, and 12. What different factors must a multiple of these numbers have?

31. Try several sets of three consecutive counting numbers to get an idea of what the divisor might be. To prove your answer is correct, use a variable. *

32. Use variables to prove that your observation is true. *

33. (a) Use the property of whole numbers that says that if $a \mid m$ and $a \mid n$, then $a \mid (m + n)$.
    (b) Notice that the numbers in part (a) are four consecutive composite numbers.

34. Use variables to represent the unknown prices and write an expression for the total cost. *

35. Solve a simpler problem: Find a 2-digit number such that when 7 is subtracted from it, the result is divisible by 7, and when 8 is subtracted from it, the result is divisible by 8. *

36. One cupcake left over each time means that if the number of cupcakes were divided by 2, 3, 4, 5, or 6, the remainder would be 1 in each case. *

37. Write the 6-digit number abc,abc in expanded form: $100,000a + 10,000b + 1000c + 100a + 10b + c$. *

38. (a) Write the palindrome $abba$ in expanded form: $1000a + 100b + 10b + a$. *
    (b) Try a few examples of palindromes with an even number of digits. Can you write a proof as in part (a)? Do you see a pattern? *

39. Each of the annual sales amounts must be a multiple of the cost of the calculator.

40. Try writing the two related numbers that are divisible by 7 in expanded form. *

*** An additional hint for this problem is given in the next section.**

41. Be sure you see how the sums are obtained. Compare the answer in one step with the sum in the next step.

42. Use the distributive property to expand the given expression. *

43. Use your calculator to show that
    11 divides 1,111,111,111,
    13 divides 111,111,111,111, and
    17 divides 1,111,111,111,111,111.

44. (a) How many 3's occur in the prime factorization of 24!?
    (b) The prime factorization of the desired factorial will contain six 3's as factors.
    (c) Remember that $12 = 2^2 \times 3$.

Section 5.2

13. Look at the method presented in this section for finding the number of different factors of any given whole number. *

14. Look at the method presented in this section for finding the number of different factors of any given whole number. *

15. For what values of n is $2^n - 1$ a prime number? NOTE: n must be a counting number.

16. GCF(24, x) = 1 means the largest whole number that is a factor of both 24 and x is 1.

17. Solve a simpler problem. Suppose there were only 10 students and 10 lockers. A list of locker numbers might be helpful. *

18. The price of a candy bar and the price of a can of pop must each be a factor of the total amount of money earned.

19. If c = price of one chicken, d = price of one duck, and g = price of one goose, then it must be true that $3c + d = 2g$. *

20. Look at the prime factorization of each number. *

21. Start with the largest possible 3-digit number.

22. Write the numbers in their expanded form, i.e., $1000a + 100b + 10b + a$. *

23. What must the sum of the primes in any row equal?

24. What is the least number of cards that could be divided exactly into 2 equal piles, 3 equal piles, or 5 equal piles?

25. Write $100a + 10b + a$ in terms of a and determine whether the resulting expression is divisible by 7.

26. Look at the prime factorization of each of the numbers.

Chapter 6
Section 6.1

19. When the fractions that work are written in simplest form, what prime factors appear in the denominators?

20. Try a few examples, substituting values for x, y, and z.

21. You will subdivide the hexagon differently for parts of this problem.

An additional hint for this problem is given in the next section.

22. Look at several other fractions of this same form. Examine the factors of the numerator and denominator in each case.

23. Try several examples in each case. If the numerator is said to be fixed, then the two fractions being compared must have the same numerators. Otherwise, you may choose any numbers you like for the two numerators.

24. What are the one-digit squares that are possible sums of the numerator and denominator?

Section 6.2

15. Remember that if a and b are whole numbers, $a < b$ if, and only if, there is a whole number n such that $a + n = b$. So, in terms of fractions, $\frac{a}{b} < \frac{c}{d}$ if, and only if, there is a fraction $\frac{n}{m}$ such that
$$\frac{a}{b} + \frac{n}{m} = \frac{c}{d} \; . \; *$$

16. Make a list of the sums with 1 term, 2 terms, 3 terms, and so on. Do you see a pattern in the answers?

17. What fraction of the restaurant do Sally and her brother together own?

18. Altogether, what fraction of his life did John spend growing up, in college, and as a teacher ? *

19. (e) Compare the sums
$$\frac{15}{50} + \frac{3}{6} \text{ and } \frac{15}{50} + \frac{1}{2} \; .$$

20. (c) What kind of numbers are 6, 28, and 496? *

21. Examine carefully how David "borrowed" in the subtraction problem.

22. Find the least common denominator by factoring all of the denominators. Then add the fractions.

23. Look carefully at the sum $\frac{1}{3} + \frac{1}{6} = \frac{1}{2}$. Notice that the least common multiple of 2 and 3 is 6.

24. If there are n terms in the numerator and denominator of the fraction, then the numerator can be written as $1 + 3 + 5 + \ldots + (2n - 1)$. *

25. Draw a regular octagon and count the number of segments that would be required to connect each vertex with every other vertex. *

Section 6.3

21. Solve a simpler problem to determine what operation is required. For example, how would you solve the problem if each load required 2 cups of detergent?

22. Use a variable. *

23. Use a variable. *

24. Try solving a simpler problem. *

25. What fraction of Kathleen's total study time was spent on subjects other than English? *

26. What fraction of the original group was still participating on July 2? *

---

* An additional hint for this problem is given in the next section.

27. Draw a diagram showing Tammy's trip to school together with the times. *

28. First determine what operation is required. In part(a), if the recipe is doubled, then twice as much flour is needed.

29. Use guess and test or use a variable. *

30. (b) What fraction of the previous year's value does the equipment have at the end of a year? *

31. Write an equation describing each of the two numbers separately and solve for the numbers. *

32. (b) How would you perform $\frac{a}{b} \div n$ using the invert-the-divisor-and-multiply approach? What would the result be?
    (c) Think of $5\frac{3}{8}$ as $5 + \frac{3}{8}$ and of $10\frac{9}{16}$ as $10 + \frac{9}{16}$.

33. In Sam's case, watch what he does with the denominators of the fractions. In Sandy's case, look at her second step.

34. You will need to use equivalent fractions. *

35. Use a variable. Let x represent the number of apples in the store.

36. Let the officers be represented by A, B, C, D, and E. Make a list of the possibilities. *

Chapter 7
Section 7.1

16. Use systematic guess and test. *

17. Let x represent the amount of money Gary had before cashing the check from Joan.

18. Let x represent the original value of the car. Then what was its value after one year?

19. Use a variable. *

Section 7.2

14. You want to calculate the time in light years. How are distance, rate, and time related?

15. What are the only possible prime factors of the denominator of a fraction in simplest form whose decimal expansion terminates? *

16. Work backward. Consider some familiar terminating decimals, for example, $0.5 = 1/2$. *

17. Beginning with the equality

    $0.\overline{1} = 1/9$, we can multiply both sides by any nonzero number we choose.

18. Same as problem 17, except begin with $0.\overline{01} = 1/99$.

19. Same as problem 18, except begin with $0.\overline{001} = 1/999$.

20. (a) Count the number of decimal places and compare with similar decimals in problems 17 - 19.
    (b) Remember that to find the fractional representation of a

---

* An additional hint for this problem is given in the next section.

number such as $0.\overline{65}$, we can start by letting $n = 0.\overline{65}$, or $n = 0.656565...$

21. (a) Look at problems 17 - 19 to determine what the denominator must be.
     (b) When might it be possible to simplify a fraction such as the one you wrote in part (a) to one that will have a decimal representation with fewer than five digits in the repetend? *

22. Write out several repetitions of the decimal and make a table. *

23. Try a simpler problem. For example, compare $1/13$ to $10/13 = 10 \times 1/13$. *

24. On an inexpensive, four-function calculator an overflow error usually occurs when the result of a calculation is too large or too small. *

Section 7.3

11. Remember to calculate the amount of juice - that is, water and concentrate.

12. You could set up a proportion like:

$$\frac{acres}{days} = \frac{acres}{days} \cdot$$

13. You could set up a proportion like:

$$\frac{peaches}{servings} = \frac{peaches}{servings} \cdot$$

14. You could set up a proportion like:

$$\frac{ounces}{weeks} = \frac{ounces}{weeks} \cdot$$

15. What fraction of each day has the man spent sleeping?

16. You could set up a proportion like:

$$\frac{\text{weight on earth}}{\text{weight on moon}}$$

$$= \frac{\text{weight on earth}}{\text{weight on moon}} \cdot$$

17. You could set up a proportion like:

$$\frac{miles}{year} = \frac{miles}{year} \cdot$$

18. You could set up a proportion like:

$$\frac{\text{altitude gained}}{\text{horizontal distance traveled}}$$

$$= \frac{\text{altitude gained}}{\text{horizontal distance traveled}} \cdot \text{*}$$

19. Ignore extraneous information in the problem and set up a proportion. *

20. (a) If the teacher: student ratio is 1:35 and there are 1400 students, how many teachers are there at this time? *
     (b) What is the total cost for all of the teachers in the school? *

21. Notice in triangles RST and XYZ that ratios of corresponding sides are equal. *

22. (a) Use the fact that the ratio of the distance from Earth to Mars to the distance from Earth to Pluto is 1:12.37.
     (b) Use the fact that Pluto is 30.67 AU from Earth, as was given in part (a). *
     (c) Use your result from part (a), which gives the AU distance from Earth to Mars.

---

* *An additional hint for this problem is given in the next section.*

23. You can solve each part by writing a proportion in a form such as one of the following:

$$\frac{\text{years}}{\text{years}} = \frac{\text{hours}}{\text{hours}} \quad \text{or} \quad \frac{\text{hours}}{\text{years}} = \frac{\text{hours}}{\text{years}}. \quad *$$

24. Let d = distance to the airport. Use distance = rate × time to solve for time. *

25. If the man had 0 dollars after the final purchase, he must have had exactly $20 before that purchase. Half of that $20 he had in his pocket and half had been given to him by his father. How much had been in his pocket before the second purchase? *

26. You could have more than one dollar's worth of change.

27. Consider the last subtraction. What number would you want to leave your opponent with so that he/she is sure to lose? What about your opponent's second-to-last subtraction? *

28. Notice that the 11-minute timer lasts 4 minutes longer than the 7-minute timer.

29. Draw a picture of the racetrack and the posts. *

30. If the ratios of successive amounts are whole numbers, each amount of money must be divisible by the preceding amount of money.

31. Make a list and look for a pattern. Compare the number of apples that must be taken with the number of apples of the same kind desired.

Section 7.4

18. (a) Remember that the value of the account is the original balance plus interest earned. *

19. How much total interest would you pay for the 15 days if you pay 0.04839% each day? *

20. 2 is what percent of 35?

21. (a) If contributions increased by 6.2% in 1991, what percent of the 1990 contributions were the contributions in 1991?

22. (b) Use your answer from part (a) to determine what percent each energy source is of the total.

23. What is the dollar amount of the discount? Is it 15% of the original price?

24. Use a variable. Let x = selling price of the car.

25. Consider this relationship: 20% of 50 is 10% of 100.

26. If you received an 8% discount on the car, what percent of the original price did you pay? *

27. (a) You want to determine what percent 8.5×10^{13} is of 5.2×10^{14}.

28. Use a variable. Let x = number of grams of protein recommended (U. S. RDA). *

29. If the slacks were made 10% longer than 40 inches, what would be their length before washing? *

* An additional hint for this problem is given in the next section.

30. Try a few examples by working the problem both ways. For example, suppose a $100 item is marked up 10% and then down 10%. What would the final price be? *

31. Suppose that Cathy has 100 baseball cards. How many do Joseph and Martin have?

32. The range is from 70% of the difference to 80% of the difference.

33. The number of outputs provided by the unit must be a whole number.

34. Try the calculation both ways. Suppose that the doctor would have charged you $100 before any discounts.

35. Use a variable or start with an arbitrary number of people. For example, suppose the population in the county was exactly 100. *

36. Simulate the game with a partner, using toothpicks, coins, etc. *

37. Try the calculation both ways. To make the arithmetic easier, suppose an item cost $100 originally.

38. Elaine will earn 3.5% interest twice a year for the next 3 years.

39. Each year she needs to earn 11% more than during the previous year.

40. Each year the value of their savings will be worth 8.25% more than the previous year. What percent of the previous year's savings is the value of their savings in any one year? *

41. What is the percent of increase from the CPI in 1985 to the CPI in 1986?

42. Make a table containing a number of examples. Compare salary, taxes and net earnings. *

Chapter 8
Section 8.1

19. (a)-(c) Try a few examples. Pay close attention to parentheses in part (c).
     (d) Remember that the identity property must hold in both directions. For example, in the case of addition, $a + 0 = a$ and $0 + a = a$.

20. Assuming that the adding-the-opposite approach works means assuming that $a - b = a + (-b)$. So this equality can be used at any time in the proof. *

21. Try several examples for each part. Be sure to also consider cases where one or both numbers are negative or zero.

22. (a) Since A is closed under subtraction, to show that a number is an element of A, you must show that it can be written as a difference using 4's and 9's.
     (b) Are there any numbers that cannot be obtained using 4's and 9's and the operation of subtraction?
     (c) Again, consider whether there are numbers that you cannot obtain using 4's and 9's.
     (d) Try some more examples as in part (c). Vary your choices of elements. *

* *An additional hint for this problem is given in the next section.*

23. Work from the bottom up.

24. Try a few examples first. To test whether the procedure will always work, write each number in expanded form, as in $10a + b$.

Section 8.2

23. What number must be in the first square of the bottom row?

24. Remember that the set of integers is an infinite set, containing all the whole numbers and their opposites. It does not, however, contain decimal numbers such as 6.9.

25. (b) First consider which cases always give a positive or always give a negative.

26. You want to calculate the number of grams in one atom of carbon.

27. (a) It might be easiest to first convert the rate of hair growth into scientific notation. *

28. How does $|x|$ compare to x if x is a negative integer?

29. See the additional properties of integers in the textbook.

30. Consider how much must be added in each case to get out of debt.

31. To get started, look at the bottom row. *

32. If $x < y$, you may multiply both sides of the inequality by a positive number and the inequality is preserved. See the Property of Less Than and Multiplication by a Positive in your textbook. *

33. There is more than one possible answer. You might make a table with columns for numbers of cows, sheep, and rabbits, as well as total cost. Try some combinations of 100 animals. *

34. If x^2 is a perfect square, then x^2 must be of the form $3n$ or $3n + 1$. *

35. Try several other examples with different integers across the top and down the left. Check the diagonal sums. *

36. If $ab = 0$, and $b \neq 0$, what can you conclude about a?

Chapter 9
Section 9.1

26. The definition of equality of rational numbers states that two rational numbers are equal if, and only if, the "cross-products" are equal. So you must show that the "cross-products" are equal in this case.

27. See section 6.2 in the textbook for an example of a proof that addition of fractions is commutative. Notice that the proof uses the fact that addition of whole numbers is commutative. Similarly, in the proofs of properties of multiplication of rational numbers, you may use any properties of integers and/or fractions.

28. (b) By the adding-the-opposite approach,
$$\frac{a}{b} - \frac{c}{d} = \frac{a}{b} + \left(-\frac{c}{d}\right).$$
Start with this definition of subtraction and show that the result you found in part (a) holds. Remember there will be two parts to the proof since the

An additional hint for this problem is given in the next section.

result is given as an if-and-only-if statement.

(c) Same as part (b) except you start by using the definition of subtraction given in part (a) and must show this means

$$\frac{a}{b} - \frac{c}{d} = \frac{a}{b} + \left(-\frac{c}{d}\right).$$

29. You may use the distributive properties of multiplication over addition for fractions and/or integers. *

30. Remember that if $\frac{a}{b} < \frac{c}{d}$, where $b > 0$ and $d > 0$, then there is a fraction $\frac{m}{n}$ such that $\frac{a}{b} + \frac{m}{n} = \frac{c}{d}$. *

Section 9.2

26. See the proof that $\sqrt{2}$ is irrational in this section of the textbook. *

27. Look at the prime factorization of 9. How does this affect the argument about prime factors of both sides?

28. Begin as you did in problem 26, by assuming that $\sqrt[3]{2}$ is rational. Then cube each side of the equation. *

29. If $\frac{a}{b}$ is a rational number, what kind of number is $\frac{a}{5b}$? *

30. (a) Just as in problem 29, assume $1 + \sqrt{3}$ is rational and then isolate $\sqrt{3}$. Show that a contradiction results.

31. Use the result proved in problem 30, part (b).

32. Try several other examples such as:
$$\sqrt{36} + \sqrt{64} \overset{?}{=} \sqrt{100}.$$
Is the result ever true?

33. For what kinds of numbers is $\sqrt{a}$ defined in this section? See the definition in the textbook.

34. Look at multiples of the Pythagorean Triple (3, 4, 5). For example, is (9, 12, 15) a Pythagorean Triple?

35. Try several pairs of values for u and v . Read the conditions for u and v carefully.

36. Use a variable. Let $x =$ smallest of the three consecutive integers. What expressions represent the other two integers? *

37. Draw a picture and label the dimensions of the rectangular region using variables. *

38. Use variables to represent the numerator and denominator of the rational number. *

39. Draw a picture. Imagine putting the two pieces of wire together and then cutting. *

Section 9.3

15. (a) Look at the graphs of the basic function types presented in this section.
    (c) Locate 100m on the vertical axis.
    (d) What physical limitations are there to this problem?

16. (b) When is $s = 90$?
    (c) When the ball hits the ground, $s(t) = 0$.

* *An additional hint for this problem is given in the next section.*

(d) What is the maximum value of s?

17. (b) What is the value of *t* in the year 1996?
    (c) Locate 8 on the vertical axis or use the graphing calculator to locate the point with a y-coordinate of 8.
    (d) Choose any point on your graph. Then find the point on the graph where the value of P is twice as large. Compare the values of *t*.

18. At what point on the route would the cyclist's speed begin to decrease? *

19. It may be helpful to sketch the graph first. *

Chapter 10
Section 10.1

14. (a) Which types of graphs can be used effectively to show the total number of each type of school? Which might best show the comparison between numbers of public and private schools of each type?

15. (a) Since the sources are listed by percent of the total budget, how might you show how much of the total each source represents?

16. (a) Which kind of graph could best show how tuition costs at each kind of college changed over time?
    (c) Which college had the greatest percent increase over the ten years?

17. (a) Which types of graphs can be used effectively to show the total number of pieces of mail received by each city?

18. (a) Which types of graphs could be used to demonstrate the change in percentages over time?

19. (a) Which types of graphs might show the relationship between the three types of funding and, at the same time, the way those relationships changed over the years?

20. (a) Which types of graphs could be used effectively to compare the percentages in each age group during each of the 2 years?

21. (a) See figure 10.2 in the textbook, for an example.
    (b) See figure 10.14 in the textbook, for an example.
    (c) What fraction of the time did exactly one head occur?

22. (b) What fraction of the time would you expect HH to occur?
    (e) See figure 10.2 in the textbook, for an example of a histogram.

23. Assume $a^2 + b^2 + c^2 = d^2 + e^2 + f^2$. Show that if the digits are rearranged, as in the second equation in the example, the resulting equation will be true. *

24. If n^2 is even, what can you say about *n*? *

Section 10.2

16. Think backward. To find the average, you divide the total points earned by the number of students

---

*** An additional hint for this problem is given in the next section.**

in the class. So what must you do to find the total number of points earned?

17. Find the total number of points scored by the original 100 students. See hint for problem 16. *

18. You have been given some unnecessary information in this problem.

19. (a)-(b) Draw a normal distribution and label it with z-scores and with percentages of data that lie in various regions. *
    (c) Think about your results from part (b). The 16th percentile corresponds to what z-score?

20. Find the mean, standard deviation, and Lora's z-score for each set of test scores.

21. (a) What will two graphs look like if they have the same mean?
    (b) What will two graphs look like if they have the same variance?

22. (a)-(b) Remember that $z = \dfrac{x - m}{s}$, where m is the mean of all scores and s is the standard deviation.
    (c) What percent of scores lie to the left of the z-score you found in part (b)?

23. (a) First find the area of each rectangle drawn in the histogram. Notice that all the rectangles have the same width. *
    (b) How do you find the median in an ordered list of scores? *

24. Be careful! The answer is *not* 90 mph. You cannot average the speeds on the two laps. *

25. (a)-(b) Draw a cube and mark the sides or draw one side of each cube and show the subdivisions.
    (c) Use exponents. Look for powers.

Chapter 11
Section 11.1

18. (a) It may be helpful to list the possible outcomes as ordered pairs. See the experiment of Example 11.2 (d) for an example.
    (b) To calculate P(D), it may be easier to use the complement of D.

19. Draw a picture similar to the first figure given in the examples. Label the cities and the distances between them in your drawing.

20. See the examples and description of the probability of a "geometric" event in problem 19. *

Section 11.2

12. Use the Fundamental Counting Property in each case. *

13. (a)-(b) See the hints for problem 12.
    (c) Use your results from parts (a) and (b).

14. (c) Compare your results from part (b) to Pascal's Triangle, as shown in Figure 11.10 in the text.

---

* *An additional hint for this problem is given in the next section.*

15. (a) Refer to Pascal's Triangle in Figure 11.10 of the text.
    (b) Compare probabilities using the table given in the example of 3 shots and your result for 4 shots in part (a).

16. (b) Which outcomes correspond to Boston winning in 2 straight games? Which correspond to Milwaukee winning after losing the first game?
    (c) If an event has several outcomes, corresponding to the ends of several branches, the probabilities at the ends of the branches should be added.

17. (a) The probability that Boston wins or Milwaukee wins is 1.
    (d) What outcome corresponds to Boston losing the second game but winning the series?
    (e) It might be easier to determine the probability that the series does *not* go for 3 games.

18. (b) To determine the probability that A wins in 4 games, consider all the possible ways in which that might occur.
    (c) One way in which A could win the series in 5 games is the outcome ABABA.
    (d) Use your results from parts (a), (b), and (c).

19. (b) Use the Fundamental Counting Property.
    (c) How many different sequences of answers are correct?
    (d) Use your answers from parts (b) and (c).

20. (a) There should be nine branches in the final row of the tree.
    (d) Use the probabilities assigned in your tree diagram, not just the number of elements in the event or sample space. *

21. What is the greatest number of *unmatched* socks you could pull out?

22. If there are three puppies, how many possibilities are there for the sexes of the puppies? That is, if each puppy is either male or female, how many elements are in the sample space? *

23. Construct a two-stage probability diagram. Notice that the probability for choosing a black ball or a white ball from box two depends on what was chosen from box one. *

Section 11.3

21. (b) For the American League team to win in four games, they must win the first and win the second and win the third and win the fourth games. Think of a tree diagram.
    (d) If the series ends in four games, it means the American League team won in four games or the National League team won in four games.
    (e) Remember that if the odds in favor of an event are $a : b$, then the probability of the event occurring is $\dfrac{a}{a + b}$.

22. (a) The probability of each sequence is the same. Think of a probability tree diagram.
    (c) If the series ends in five games, it means the American League team won in five games or the National League team won in five games.

An additional hint for this problem is given in the next section.

23. (a) The probability of any sequence of 4 A's and 2 N's is the same. Think of the probability tree diagram described in the problem.
    (c) If the series ends in six games, it means the American League team won in six games or the National League team won in six games.

24. (a) Same as the hints for problem 23 except the sequences will consist of 4 A's and 3 N's.

25. (b) Remember that if the odds in favor of an event are $a : b$, then the probability of the event occurring is $\frac{a}{a+b}$.
    (c) Remember that the expected value of an experiment is the sum
    $$E = v_1 p_1 + v_2 p_2 + \ldots + v_n p_n .$$
    In this problem v_i = number of games in the series and p_i = probability of the series ending in that many games.

26. You can simulate the experiment by rolling a die. Use a different number on the die for each prize. Ask someone to work with you on this simulation. Keep a careful tally of the numbers that show up on the die and of the number of rolls required in each trial. *

Chapter 12
Section 12.1

12. In each case analyze what has been done to the first picture in order to obtain the second.

13. Use a ruler and/or protractor to attempt to draw the parallel or

perpendicular lines on triangular dot paper.

14. Copy the triangular lattice on your paper. Using a ruler and/or protractor, draw parallel or perpendicular sides of quadrilaterals as needed. Be sure to check the lengths of sides. *

15. Examine one row of the lattice at a time. How many figures of each type can be drawn, for example, using as vertices the points in the first row? *

16. In each part you might find it helpful to start drawing from the center of the hexagon. *

Section 12.2

8. Make drawings of the possibilities, being systematic in your attempts. You might find graph paper useful here. *

9. Try cutting the shapes out of graph paper and fitting them together. Or you might draw a 5 by 8 rectangle on graph paper and try to fill it with tetrominos.

10. In each case, consider whether it is possible for a triangle to be of both types. If so, is one type always of the other type? For example, in part (a), ask yourself, "Is it possible for a triangle to be both isosceles and scalene? If so, is a scalene triangle always isosceles, or is an isosceles triangle always scalene?"

11. You may trace the whole figure, including both diagonals, to show that the diagonals are congruent. Try to match one diagonal of the

An additional hint for this problem is given in the next section.

tracing with the other diagonal of the original.

12. How much of a turn must you make with the tracing paper before the traced image coincides with the original square?

13. See the hint for problem 12.

14. (a) If you fold the kite on a diagonal, do opposite angles coincide?
    (b) How is a rhombus related to a kite?

15. If you fold the rhombus on a diagonal, do the two portions of the other diagonal coincide?

16. How can you show that two lines are perpendicular by folding? You might try a few lines first.

17. (a) What lengths are possible for the two congruent sides?
    (b) What lengths are possible for the longest side?
    (c) How long can each side of the triangle be?

Section 12.3

8. You might use variables to represent the unknown measures of the angles. *

9. Use a variable.

10. What angles are congruent to –1? Use the parallel lines and vertical angles. *

11. You may use the Corresponding Angles Property. That is, if congruent corresponding angles are formed, then the lines are parallel. *

12. (a)-(b) Use vertical angles and the Corresponding-Angles Property.

13. Use a variable.

14. The Corresponding Angles Property states that if lines are cut by a transversal so that corresponding angles are congruent, then the lines are parallel. What are some pairs of corresponding angles? *

15. (b) Can you draw an example with four points of intersection? With five points of intersection?

16. (b) What is the maximum number of points of intersection that a circle can have with one side of a triangle?

17. Make a table and look for a pattern. *

Section 12.4

16. Try a simpler problem. What if there were only 4 people in the room? Only 5 people? *

17. Use the fact that the sum of the angle measures in a triangle is 180°. Look at both large and small triangles in the figure. *

18. What is the measure of each angle of the regular pentagon in the center of the star? *

Section 12.5

24. You might look at the shape on top of the cube. What side would be adjacent to it if the figure at the left were folded? Consider the orientation, too. You might also

An additional hint for this problem is given in the next section.

find it helpful to construct a paper model and fold it to form a cube.

25. Use a piece of clay, a block, a die, or a sugar cube as a model. Not all the planes of symmetry are horizontal or vertical.

26. Notice that the axis of rotational symmetry shown passes through two opposite faces of the cube. You may find a block, a die, a sugar cube, or other model of a cube helpful.

27. (a) Try turning a block, a die, or other model of a cube to check this out. Mark the top face, if necessary.
    (b) How many pairs of opposite vertices are there in a cube?

28. (a) Use a block, a die, or other model of a cube to try rotating on this axis. Mark the top face, if necessary.
    (b) How many pairs of opposite edges are there in a cube?

29. Cut the cardboard center of a paper towel roll along the lines and see what shape results.

30. (b) Which polygons will tessellate the plane?
    (c) If you have models of regular polyhedra available, try using sets of them to surround a point in space. There is at least one regular polyhedron that will tessellate space by itself. Are there others?

Chapter 13
Section 13.1

26. Use dimensional analysis. You can form a unit fraction using the fact that 1 gallon of water weighs about 8.3 pounds.

27. Use dimensional analysis. How many microliters of blood are in one liter?

28. Use dimensional analysis to perform the necessary conversion in each part of the problem.

29. See the descriptions of portability, convertibility, and interrelatedness in your textbook. *

30. (a) Use dimensional analysis to convert from miles per second to miles per year.

31. Since both trains are moving at 50 mph, the resultant rate is actually 100 mph. *

32. (a) How many square feet are in one acre? Use dimensional analysis to convert from acres to square feet.
    (b) Use dimensional analysis to convert from cubic feet to pounds.
    (c) Use dimensional analysis to convert from pounds to gallons.

33. (a) Draw a new ruler with marks at only 1, 4, and 6 units. How can you measure with it?
    (b)-(c) Consider the new ruler you made for part (a). How might a similar ruler be constructed for each of these two problems?

An additional hint for this problem is given in the next section.

34. (a) First determine the number of cubic feet in a cord of wood. *
    (b) How much money would the son make in one day?
    (c) How many cubic feet of wood are cut in one day?
    (d) You can solve this problem using a proportion:
    $$\frac{\$85}{100 \text{ ft}^3} = \frac{?}{?} \cdot *$$

35. How does the time required to hike uphill compare to the time required to hike downhill? *

Section 13.2

25. Think of a 3-4-5 right triangle.

26. Express the area of the enclosure in terms of x and y. *

27. Visualize the hypotenuse of a right triangle.

28. The area of the large figure equals the sum of the areas of its parts. *

29. Use the Pythagorean theorem.

30. Review your solution to problem 29.

31. Use the Pythagorean theorem and/or the formula for the area of a triangle.

32. (a) Make a right triangle having the side of length 5 as its hypotenuse.

33. Notice that the sum of the lengths of the bases of the eight triangles is the perimeter of the octagon.

34. Draw pictures and use the result in problem 33(c).

35. Check to see if $a^2 + b^2 = c^2$.

36. (b) Determine the number of boundary points, b, and the number of inside points, i.

37. Recall that $C = 2\pi r$.

38. Use a variable for the width of the rectangle. What variable expression would represent the length of the rectangle?

39. Is it possible to draw a triangle with two sides congruent? With all three sides congruent?

40. Think about the Pythagorean Theorem. If a triangle has a right angle, then it must be true that $a^2 + b^2 = c^2$, where a, b, and c are the lengths of the sides of the triangle. *

41. If $a^2 + b^2 = c^2$, then a triangle with sides of lengths a, b, and c is a right triangle. What type of triangle is it if $a^2 + b^2 > c^2$?

42. (a) What must be true about the sums of the lengths of the sides of a triangle?
    (b)-(d) See the hint for problem 41.

43. Remember that Hero's formula, as given in exercise 17 of this section, uses the lengths of the sides of a triangle to find the area of a triangle. *

44. (a) What kind of angles are the 7.5° angle and the angle at the center of the earth?
    (b) You can use a proportion here:
    $$\frac{500 \text{ mi}}{7.5°} = \frac{?}{?} \cdot$$

An additional hint for this problem is given in the next section.

45. The diameter of the hole is also the diagonal of the square. Use a variable to find the length of the sides of the square.

46. Draw a right triangle in the figure by connecting the center of the smaller circle to point B. Label the lengths of the sides of the triangle using a variable. *

47. What is the total area of all four semicircles in the drawing? What is the total area of the square? *

48. Find the total area inside the circle(s) and inside the square. *

49. If the side of an equilateral triangle measures one unit, what is the height of the triangle? *

50. In each case the distance from Portland to the Aral Sea is a fraction of the circumference of a circle. What are the two circles and what fraction of the circumference is involved in each case? *

51. (a) Recall that the formula for the area of a circle is πr^2.
    (b) Use the area of the inner circle found in part (a).

52. Use a variable to represent the width of one ring.

Section 13.3

8. (a) Each regular hexagon can be subdivided into six equilateral triangles.

9. Use variables or choose any dimensions you like for the original box. Then determine its surface area. What happens to the surface

area of the box if you double those dimensions?

10. Use wooden cubes, plastic cubes, or sugar cubes, if available. *

11. Use dimensional analysis. Remember that surface area is measured in square units, so the scale must be adjusted.

12. Draw a picture and label the sides of the box with the area measures. Represent the lengths of the edges of the box using variables. *

13. Remember that for a sphere, the volume is given by $V = \frac{4}{3}\pi r^3$ and the surface area is given by $S = 4\pi r^2$. *

14. Use a variable or choose any radius for the sphere. Then calculate its surface area. What happens to the surface area when you reduce the radius by half?

15. What is the circumference of the cylinder?

16. The sphere must touch the top, bottom, and sides of the cylinder. *

17. Try forming a "barber pole" with red, white, and blue stripes out of a piece of paper. Roll your paper into a cylinder. What do the stripes look like when the paper is unfolded? *

Section 13.4

5. Use wooden cubes, plastic cubes, or sugar cubes, if available. When you are calculating the surface area, remember to count only the exposed faces.

---

* *An additional hint for this problem is given in the next section.*

6. Each shape has a cut made in it. Find the volume of the figure without the cut.

7. (a) Calculate the area of each face.
    (b) How would you name the shape of the pool?

8. (a) Use a variable to represent the radius of one tennis ball. How might you represent the height of the can?
    (b) First find the volume of the can and of the tennis balls.

9. How do you calculate the volume of any prism?

10. The Great Wall of China forms a giant prism. *

11. Find the volume of the cylinder.

12. How do you calculate the volume and lateral surface area of any pyramid?

13. (a) Each triangle has two sides of length 16 cm. To what dimension of the pyramid do these sides correspond?
    (b) How do you calculate the area of a regular hexagon?

14. (b) Use dimensional analysis. Remember that volume is measured in cubic units, so the scale must be adjusted.

15. (a) To find the area of a regular hexagon, you can subdivide it into six equilateral triangles.
    (b) Remember that the volume of a prism is given by $V = Ah$. *

16. (a) If the inside diameter is 60 feet, what is the capacity of the spherical tank?

17. What is the inside radius of the sphere? *

18. (a) By how much did the volume of water in the aquarium increase? *
    (b) What is the total volume of the marbles?

19. Draw a picture of a right square prism, labeling the dimensions. Write expressions representing the volume and surface area. Do the same for a new prism in which all the dimensions have been doubled.

20. Which arrangement provides the largest opening for water to pass through?

21. (b) Use the formula for the volume of a cube and solve for s, where s = length of an edge of the cube.

22. (a) Assume the height is unchanged. Try an example or use variables for the radius and height of the cylinder.

23. Draw a picture and label the sides of the box with the area measures. Use variables to represent the lengths of the edges of the box. *

24. How much larger is the big circle cut by the post hole digger than the small circle?

25. First find the volume of the original tank.

26. Be sure that you have the correct dimensions for each piece of lumber. That is, are the dimensions exact?

* An additional hint for this problem is given in the next section.

27. Find the volume of water required to irrigate the field. Be sure that your units are consistent.

Chapter 14
Section 14.1

17. (a) What distances did the hikers step off to be equal? *
    (b) Which congruence property holds?
    (c) Use congruent triangles. *

18. You might make models with straws and tacks similar to the frameworks shown and then test them to see which ones are rigid.

19. Refer to problem number 18.

20. (a) Try to draw two noncongruent triangles with three corresponding angles congruent or with two sides and an angle congruent.
    (b)-(d) Consider the possibilities. For example, with four corresponding parts congruent, you might have three sides and one angle. What would that tell you about the two triangles?

21. If two corresponding angles of two triangles are congruent, what can be said about the third angles?

22. (b) Use the fact that you were given $\angle ABC \cong \angle WXY$. Also, from part (a) you know that $\angle 2 \cong \angle 6$.
    (c) From part (b) and the given information, you have two corresponding parts congruent. What other congruent parts would show the triangles congruent?

(d) Use the same kind of argument as you used for part (b).

23. Use a variable. Let $m(\angle D) = x$. Then what could be used to represent $m(\angle E)$?

Section 14.2

7. Form a proportion using the ratio of height to waist measurement.

8. (a)-(b) Form a proportion using the distance from the projector and the thumb height.

9. Draw a picture. Remember that the triangle will be at eye level and lines up with the point on the trunk at which you will cut the tree. *

10. Draw a picture. What triangles are similar? *

11. (a) What conditions must be satisfied for a quadrilateral to be a parallelogram?
    (b) Use your result from part (a).
    (c) Use your result from part (b).

12. (a) What similar triangles are formed by the parallel lines and transversals?
    (b) Write a proportion using corresponding sides of $\triangle PAD$ and $\triangle PBE$.
    (c) Write a proportion using corresponding sides of $\triangle PAD$ and $\triangle PCF$.
    (d) Use your results from parts (b) and (c).
    (e) Use your results from parts (b) and (d).

---

* *An additional hint for this problem is given in the next section.*

13. (a) What segments on transversals *n* and *o* correspond to $\overline{AB}$ and $\overline{BC}$?

14. First find CE . *

15. What triangles are similar in the drawing? What corresponding sides include the segment $\overline{BD}$?

16. (b) What volume of water will be in the cup when it is half full? *

17. Use a variable. Let AP = *x*. Then what expression can be used to represent PB?

18. Try several examples of pairs of right triangles. Compare the areas of the triangles. *

19. There are three pairs of similar triangles contained in the drawing. Can you name them? Which pair yields a proportion involving the sides with lengths *x*, *a*, and 1?

20. (d) Make a table and examine what happens to one side of the original triangle as *n* increases. *

21. (d) Make a table and examine what happens to the area added on to one side of the original triangle as the value of *n* increases. *

22. There are three right triangles in the figure. Use the AA Similarity Property to show that the triangles are similar. *

Section 14.3

13. Draw medians and angle bisectors in several types of triangles--

scalene triangles, isosceles triangles, etc.

14. See the hint for problem 13.

15. Construct the perpendicular bisector of a segment $\overline{AB}$ following the procedure given in the textbook. Note the radii used for your arcs. Label the points of intersection as in the given figure. Draw in segments $\overline{AP}$, $\overline{BP}$, $\overline{AQ}$, and $\overline{BQ}$. *

16. The bisector of ∠A divides the triangle into two triangles. By what congruence property will the two triangles be congruent? *

Section 14.4

12. (c) Does the circumcenter lie inside, outside, or on a side of the triangle in each case?

13. (c) See the hint for problem 12.

14. (c) See the hint for problem 12.

15. Think of the edge of the lined paper as a transversal. The parallel lines on the page intercept congruent segments on that transversal. *

16. The midpoint of a square is the point of intersection of its diagonals. *

17. Examine the construction given in Example 14.9 in the textbook. Notice that if PQ = *a* and PS = *b*, then PT = *ab*. If the product of the lengths (or *ab*) and the length of one of the segments (say *a*) were known, then the other length (*b*) could be constructed. *

*** An additional hint for this problem is given in the next section.**

18. (a) If $\dfrac{a}{x} = \dfrac{x}{b}$, $a = 1$, and $b = 2$, then
         what is the value of x?
     (b) What segments in the drawing
         or in your construction are
         congruent by construction?
         What triangles in the drawing
         are congruent? *

Section 14.5

1. (b) What do you know about the
        base angles of an isosceles
        triangle?
    (f) What is the definition of the
        perpendicular bisector of a line
        segment?

2. (a) What is the definition of a
        rhombus?
    (e) What is the definition of an
        angle bisector?

3. (a) Remember that a rhombus has
        all the properties of a
        parallelogram.
    (b) You will probably want to use
        the result you proved in
        problem 2.

4. Draw a pair of parallel lines cut by
    a transversal. Which angles formed
    are congruent and which are
    supplementary? *

5. Use the result proved in problem 4.

6. Draw quadrilateral ABCD and
    mark congruent sides. After you
    have drawn in diagonal $\overline{BD}$, what
    triangles can be shown to be
    congruent? *

7. What is the sum of the measures of
    the angles of a quadrilateral? You
    might want to use a variable here.
    Draw a quadrilateral PQRS, mark

congruent angles, and label their
measures using variables. *

8. Show that $\triangle ABE \cong \triangle CDE$. *

9. What is the definition of a square?
    That is, what must be shown to
    verify that STUV is a square? *

10. Use congruent triangles to show
    that $\angle STU \cong \angle TSV$. *

11. Since the diagonal of a
    parallelogram cuts it into two
    triangles, you can complete part of
    the construction of the
    parallelogram by using the SSS
    construction for a triangle.

12. See the hint for problem 11.

13. You might try sketching an
    isosceles trapezoid first. Label the
    bases and the legs of the trapezoid.
    If you draw perpendiculars from
    the endpoints of the shorter base to
    the longer base, what do you notice
    about the triangles formed at each
    end of the trapezoid? *

14. What does it mean for AC to be the
    geometric mean of AD and AB?
    How can this fact be stated in
    terms of ratios of corresponding
    sides of triangles? *

15. (a)-(c) What additional conditions
         must be met for a
         parallelogram to be a
         rectangle, rhombus, or
         square? How do those
         conditions affect the
         diagonals of PQRS?

16. Identify congruent angles in the
    drawing in order to determine
    which pairs of corresponding sides

---

* An additional hint for this problem is given in the next section.

are referred to in the proportions given, such as, $\dfrac{b}{x} = \dfrac{x+y}{b}$. *

17. You are given that $\triangle ABC$ and $\triangle A'B'C'$ have three pairs of corresponding sides congruent . But, you must show that the triangles are congruent without using the SSS congruence property. *

18. Assume that $\dfrac{AC}{AB} \neq \dfrac{A'C'}{A'B'}$. As stated, there must be a point D' on $\overline{A'C'}$ where $\dfrac{AC}{AB} = \dfrac{A'D'}{A'B'}$. What triangles must then be similar by the SAS similarity property? Remember you were given that $\angle A \cong \angle A'$ and $\angle B \cong \angle B'$. *

Chapter 15
Section 15.1

21. Try holding a pencil and revolving it about a fixed axis to get an idea of the shape that results in each case. For example, in part (a), hold the pencil parallel to the line you will revolve it about. *

22. Draw x-, y- and z-axes as shown. Mark units on each of your axes. If the x-coordinate of a point is positive, move forward that many units along the x-axis. If the y-coordinate is positive, move right.

23. It might help to mark units on each of the three axes or to label the ends of the axes as positive or negative.

24. You might find it helpful here to label the ends of each axis as positive or negative.

25. (a) Notice that the x-coordinate of R will be the same as the x-coordinate of point P.
    (b) Use the coordinate distance formula for two dimensions since the x-coordinates of P and R are the same.
    (d) Use Pythagorean Theorem here since $\overline{PQ}$ is the hypotenuse of right triangle $\triangle PQR$.

26. Remember that if P, M, and Q are collinear with M between P and Q, then PM + MQ = PQ. Use the distance formula and the given coordinates for P, M, and Q to show that this relationship holds.

27. If (x, y) lies on the line l, then the slope of the line segment between (x, y) and any other point on line l must be m.

28. If you assume that the two lines intersect, then they must have a point in common. Use this point together with one point on each of the two lines to see if you can arrive at a contradiction. *

Section 15.2

32. Sketch x- and y-axes. Remember that the initial side of the angle lies on the positive x-axis; that is, you will measure the angle starting from the positive x-axis.

33. You may find it convenient for graphing here to draw the axes and then to draw additional lines through the origin at "convenient" angles such as $45°$, $60°$, $120°$, etc.

***

* An additional hint for this problem is given in the next section.

34. Note that the distance from P to D will be $\sqrt{x^2 + (b - y)^2}$. You need to find the value of this expression. You do NOT need to determine the values of x, y, or b. *

35. (b) Look at your last entry in the table of part (a).
    (c)-(d) Remember if the equation of a line is written as $y = mx + b$, the slope is the coefficient of x, or m.

36. What do you know about the slopes and y-intercepts of the two lines if the system of two linear equations has no solution, exactly one solution, or infinitely many solutions?

Section 15.3

8. Use the coordinate distance formula and the given coordinates to show the lengths of the diagonals are the same.

9. (a) To find the equation of the line containing the median from the vertex A, first find the midpoint of side $\overline{CB}$. *
   (d) Solve simultaneous equations.

10. (a) To find the length of the median from A, find the distance from A to the midpoint of $\overline{CB}$. Be sure to write all distances in simplest radical form.

11. (a) Write AB = BC using the distance formula and the given coordinates for points A, B, and C.
    (b) Find the slopes of $\overline{AC}$ and the median from vertex B. *

12. (a) Remember that the altitude from vertex R will be perpendicular to side $\overline{ST}$. What does this mean in terms of slope? *
    (d) Solve simultaneous equations.

13. What does it mean for the diagonals of a parallelogram to "bisect each other"? *

14. (a) To show $\overline{MN} \parallel \overline{AC}$. what must be shown about the slopes of the two line segments?
    (b) Use the distance formula and the given coordinates to determine MN and AC.

15. First find the coordinates of M, N, O, and P. *

16. How can you use slopes to show that the diagonals are perpendicular?

17. Use the distance formula to show the two medians are the same length.

18. (a)-(c) Remember that if a line has a slope of m, where $m \neq 0$, any line perpendicular to it must have a slope of $-1/m$.
    (b) Solve simultaneous equations.

19. Consider the possible lengths of the horizontal side. *

20. Try using a variable to represent each of the girls' ages.

21. Try using variables. For example, let x = number of bicycles that passed the house.

---

* An additional hint for this problem is given in the next section.

22. Try using a variable to represent Mike's investment and a variable to represent Joan's investment.

23. Try letting x = number of quarters spent and y = number of dimes spent.

24. Try a simpler problem. Examine a smaller grid or look at a portion of the grid shown. *
    (c) Use your results from (a) and (b).

25. It may be easier to count the squares if for each different size of square you count the number of possible locations for one vertex of the square. *

26. Draw several examples and look for a pattern in the relationship between the numbers of blacks and whites. *

Chapter 16
Section 16.1

35. Under the translation T, some point was translated upward and to the right to have an image of A. Where must that point have been originally?

36. The translation T moves each point down one unit and to the right two units.

37. (a) Connect each lattice point on l in turn to point O. Then locate the image point.
    (b) Use a new lattice for part (b). The reflection line is the vertical line. How does the line of reflection relate to a point and its image?

38. What will each type of transformation will do to point A? *

39. Not all of these are possible. *

40. Draw coordinate axes and plot several random points. Connect the points to form a polygon. Using the given formula, determine the image of each point and connect the vertices to form the image of the original polygon. Can you see the reflecting line? *

41. Examine what happens to the x-coordinates and y-coordinates separately. Do you see a pattern or can you write a formula to describe what happens? *

Section 16.2

17. (a) See the argument associated with figure 16.26 in the textbook.
    (b) Think about how to show that three points are collinear using slopes. This argument will be similar.

18. See the argument in the textbook that isometries preserve angle measure.

19. Remember that rotations are isometries and that isometries preserve angle measure. See the argument in the textbook. *

20. Remember that points A, B, and C are collinear and point B is between points A and B if, and only if, AB + BC = AC.

21. Sketch lines p' and q' and then draw a transversal that intersects them. Which angles must be congruent to angles 1 and 2? *

An additional hint for this problem is given in the next section.

22. See the argument associated with Figure 16.30 of the textbook, which demonstrates that $\triangle PQR \cong \triangle P'Q'R'$.

23. Draw several examples for each case, if possible. Note angles and centers of rotation, lines of reflection, etc. *

24. How do you locate the center of the size transformation? The figures showing size transformations in this section of the textbook might be helpful here.

25. First find the center of the size transformation, point P. *

26. Refer to your work for problem 25. For this problem, it is probably easier not to find the center of magnification first. *

27. What segments are congruent if l is the line of reflection? Can you use those segments to show a pair of triangles congruent? *

Section 16.3

11. Let M be the midpoint of diagonal $\overline{AC}$. What are the images of points A, C, and D, under the half-turn centered at M? *

12. Use corresponding parts of congruent triangles.

13. What is the image of point A under the reflection M_{BP}?

14. (a) Draw in the diagonals of the kite. Look at one of the triangles formed and use the result from problem 13.

(b) What does it mean for a polygon to have reflection symmetry? *

15. What is the image of point B under the reflection M_{AP}?

16. (a) What is the image of A' under the reflection M_S? How does this point relate to point A? *
    (b) Use the variables x and y. *

17. (a) Choose two intersecting sides, say the bottom and the right-hand side. Reflect B across the right-hand side to a point B'. Then reflect B' across the extension of the bottom of the pool table.
    (b) Refer to Example 16.14. *

18. What size transformation could be used to "shrink" square ABCD so that its image A'B'C'D' is congruent to EFGH? *

19. Use clay, a die, or paper cube as a model. Mark the midpoints A, B, C, and D. *

---

* **An additional hint for this problem is given in the next section.**

ADDITIONAL HINTS - PART A PROBLEMS

Chapter 1
Section 1.1

5. Remember that you might not use all of the symbols.

9. The stools need not be evenly spaced along the walls.

14. You can also use the guess and test strategy here.

17. Notice that the inner two squares touch more squares than squares in any other position. What might you conclude about which digits to place there?

18. Remember that the large figure is a rectangle, and the opposite sides of a rectangle have the same length.

19. Let s = son's current age. Then his age seven years ago was $s - 7$.

20. If the man's age at death was $\frac{1}{29}$ of his birth year, the year of his birth must be divisible by 29.

21. What values could the letter U have?

22. You can use a variable to represent the necessary increase in her hourly wage.

24. You might find a diagram useful here. Try drawing line segments to represent time on each of the two timers.

26. (b) By how much can the hundreds digits of the two numbers differ? Which digits in the tens and units place

will make the two numbers have close to the same value?

28. Try using the first few terms to get a sum that differs from 100 by only 1 or 2.

Section 1.2

2. How does each term compare to the previous one? Is there a common difference or a common ratio?

9. See the formula developed in problem 6(f).

16. For each column, consider the difference between every other number.

24. (d) What do all but the first term have in common?

25. In attempting to write a positive integer as a sum of 9's and 4's, you might first consider how many 9's could be in the sum.

Chapter 2
Section 2.1

25. How many different elements of B could be paired with one in A?

26. Make a table comparing the number of elements in the set to the number of subsets and look for a pattern.

31. Start with the innermost region and be sure not to count any persons twice.

33. Try drawing line segments connecting some points on $\overline{AC}$ with points on $\overline{AB}$.

Section 2.3

20. Let the number = $10a + b$, where a and b are the tens and units digits.

22. More than one answer is possible for (d).

Section 2.4

16. (c) Try thinking of the number of triangles formed as a sum, rather than as a single number. That is, in step 2, think of $1 + 2$, rather than 3.

Chapter 3
Section 3.1

18. Choose a row, column, or diagonal containing two numbers. What must be the sum of the two missing numbers? Look for combinations of unused numbers with this sum.

20. See if you can find a relationship between the sum of the two digits in a number and the number of steps required to obtain a palindrome.

Section 3.2

18. (a) Remember that each row sum is the same. Once you know what that sum is, use guess and test to place the digits.
    (b) Remember that when you multiply powers together, you add the exponents.

23. In this problem there are actually three unknowns: the number of cups of tea per person, the number of cakes per person, and the number of persons in the group.

24. To get started, suppose that all five numbers were exactly the same. What would each number be?

Section 3.3

12. What kind of sequence do the candy bar prices form?

14. How can you relate the first term on the left-hand side of the equation to the first term on the right-hand side?

15. You might name your four toppings and consider that for each topping, it either is or is not included on a given pizza.

17. Try drawing pictures of three sets A, B, and C such that $n(A) < n(B)$ and $n(B) < n(C)$. Indicate the related 1-1 correspondences.

18. How would the difference be expressed in terms of a and b? Does this expression factor?

19. (a) What happens if you add c to both sides of this equation?

Chapter 4
Section 4.1

28. Consider the magnitude of the factors and look at the final digit in the product.

32. To verify the result in general, represent the number as $10a + 5$ and then square.

39. Look for a pattern in the kinds of products for which this shortcut works. How are the units digits related?

40. Be sure to specify what happens to every digit of the number when you round using this method.

Section 4.2

23. Notice that the sum of the first two digits must be 9 or less.

29. "Carries" will be required in each column so that the sums in each column will differ.

Section 4.3

9. Imagine base pieces: units, longs, and flats.

Chapter 5
Section 5.1

14. See the justification for divisibility by 2 in this section of the text.

15. Examine what factors appear on the right-hand side. How do they compare with the number you are testing as a divisor, or with the factors of that number?

20. Look for a pattern in the relationship between the number of 1's in the number and its factors.

24. Be sure to also use the quick mental tests for divisibility. For example, you know 115 will not be one of a pair of twin primes since it ends in a "5" and so is divisible by 5.

31. Let n = the first of the three consecutive counting numbers. Then what expressions would represent the next two counting numbers? next three numbers?

32. Let a = the first number and b = the second number. Then the sequence of numbers will look like this: $a, b, a + b, a + 2b, \ldots$ What is the seventh number? What is the sum of all ten numbers?

34. What factor do all the terms in your expression have in common?

35. If you subtract 7 from a number and the result is divisible by 7, then the original number must have been divisible by 7.

36. If 1 were subtracted from the number of cupcakes, the result would be divisible by 2, 3, 4, 5, and 6.

37. Combine like terms and factor.

38. (a) Collect like terms and factor.
    (b) Which of the following palindromes are divisible by 11: 1001, 10001, 100001, 1000001, 10000001? Can you express any palindrome with an even number of digits as a sum of terms involving known multiples of 11?

40. If 7 divides 1001, then 7 will divide any number or expression that has 1001 as a factor.

42. Look for terms that have a common factor of 11.

Section 5.2

13. Remember you are to find the smallest number with a certain number of factors. So you want to

choose the smallest possible prime factors and the fewest of them.

14. To characterize the sets of numbers, look at the prime factorization of the numbers.

17. How many openings and closings can a locker have if it is to remain open at the end?

19. Use systematic guess and test to find the solution. There is only one whole-number solution to the problem. (Remember that the price cannot be a fraction of a dollar nor can it be negative.)

20. What do you know about the prime factors of a perfect square?

22. Find the sum of the two numbers and combine like terms.

Chapter 6
Section 6.2

15. Make substitutions, where possible, to try to show that $\frac{a}{b} < \frac{e}{f}$, etc., by demonstrating that a fraction added to $\frac{a}{b}$ equals $\frac{e}{f}$.

18. Let t = number of years John lived.

20. See problems 10 and 15 in Section 5.2A of the text.

24. Compare the last term in the denominator to n.

25. Another approach might be to solve a simpler problem. Suppose there were only four players. Name them and list all the pairings that would be required.

Section 6.3

22. Let x = the number of barrels of oil used for transportation in the U.S.

23. Let x = the total number of students enrolled.

24. Suppose that each gallon of water required exactly two ounces of concentrate. What operation would you perform to determine the number of gallons of mix that could be made?

25. What fraction of her study time does Kathleen spend on Spanish?

26. Use a variable. Let x = the number of employees originally enrolled in the fitness program.

27. What fraction of the trip does Tammy make between the grocery store and the bicycle shop?

29. To use a variable, try letting x = the total number of games played.

30. (b) Calculate the value of the equipment for each of several years until the value is less than $40,000.

31. If n = the first number, then the first number divided by one more than itself will be represented by the expression $\frac{n}{n+1}$.

34. There is more than one correct answer for each part.

36. Remember that selecting A and B is the same as selecting B and A.

Chapter 7
Section 7.1

16. What must the sum in each row be?

19. Let x = the number of candy bars Todd ate on the first day.

Section 7.2

15. If a number is divisible by only 2 or 5, what digits must it end in?

16. Write the prime factorization of the denominators of some fractions in simplest form whose decimal representations terminate. Look for a pattern.

21. If a decimal has 1, 2, 3, or 4 digits in the repetend, what must the denominator of the corresponding fraction be when it is completely simplified?

22. Make a table listing which digit appears in which position. For example, for 1/13:

Position of digit	digit
First	0
Second	7
Third	6
.	.
.	.
.	.

Look for a pattern in when the 6's appear, etc.

23. What happens when you multiply a decimal by 10 or 100?

24. Think of how you multiply fractions: $\frac{a}{b} \times \frac{c}{d} = \frac{ac}{bd}$.

Section 7.3

18. Be sure to be consistent with the units in the problem.

19. You could use a proportion like

$$\frac{\text{length of model}}{\text{wingspan of model}} = \frac{\text{length of plane}}{\text{wingspan of plane}}.$$

20. (a) How many teachers would be needed for a teacher : student ratio of 1: 20?
     (b) Cost per pupil = (total cost) ÷ (number of pupils).

21. In the case of triangles $\triangle RST$ and $\triangle XYZ$ we could say that $\frac{RS}{XY} = \frac{ST}{YZ}$.

22. (b) You can write a proportion in a form such as one of the following:

$$\frac{\text{AU distance}}{\text{AU distance}} = \frac{\text{distance in miles}}{\text{distance in miles}}$$

or

$$\frac{\text{AU distance}}{\text{distance in miles}} = \frac{\text{AU distance}}{\text{distance in miles}}.$$

23. (d)-(f) Use your results from parts (a) - (c). For example, you know the number of years corresponding to one second. So, for part (e), you could write:

$$\frac{\text{seconds}}{\text{years}} = \frac{\text{seconds}}{\text{years}}.$$

You might find a time line labeled in both years and hours helpful.

24. How does the time required for the
     trip at 60 mph compare to the time
     required for the trip at 30 mph?

25. You could also solve this problem
     by using a variable. Let $x =$
     amount of money in the man's
     pocket when he entered the store.
     Work forward in this case.

27. You might find it helpful to solve a
     simpler problem first. Suppose you
     started with 20 or 30 instead of
     100.

29. How many lengths between posts
     have been covered in the 8
     seconds?

Section 7.4

18. Use a variable to determine the
     amount of the original investment.

19. If you pay 15 days before the due
     date, you pay no interest.

26. Use a variable. Let $x =$ the original
     price of the car.

28. If 3 grams of protein is 4% of the
     U. S. RDA, how many grams are
     recommended per day?

29. Once the slacks were made longer,
     what would be the length of the
     slacks after washing, assuming
     that they shrink by 10%?

30. If you want to verify your result in
     general, use a variable. Let $x =$ the
     initial price of an item.

35. If the population increased by 2.8%
     in 1991, what percent of the
     population in 1990 was the 1991
     population?

36. Try going first and going second.
     What configurations do you want to
     leave your partner in order to win?

40. How many times will their savings
     increase by 8.25%?

42. Is there a salary that gives a
     maximum net earnings?

Chapter 8
Section 8.1

20. To show that the missing-addend
     approach holds, you must show
     two things:
     (1) If $a - b = c$, then $a = b + c$
     (2) If $a = b + c$, then $a - b = c$.

22. (d) What relationship do you see
          between the set A and the two
          numbers that were given as
          elements of A?

Section 8.2

27. (a) How many seconds are there
          in one month?

31. Remember that all rows, columns,
     and diagonals must have the same
     product.

32. Notice that the problem states that
     both x and y are positive integers.

33. What is the maximum number of
     cows that could be purchased with
     $1000?

34. Each whole number x must be of
     the form $3n$, $3n + 1$, or $3n + 2$. Try
     each possibility to see what
     x^2 looks like.

35. To check whether this procedure will always work, use variables. Let x, y, and z be the numbers across the top and a, b, and c be the numbers down the left side.

Chapter 9
Section 9.1

29. Begin with the left-hand side of the equality and show it is equal to the right-hand side.

30. Start by assuming $\frac{a}{b} < \frac{c}{d}$, and try to show that $\frac{a}{b} + \frac{e}{f} < \frac{c}{d} + \frac{e}{f}$.

38. Use algebra to find the sum of the number and its reciprocal. What do you know about the denominator of the sum if it is an integer?

Section 9.2

26. Start by assuming that $\sqrt{3}$ is rational, i. e., that there is a rational number $\frac{a}{b} = \sqrt{3}$. Then square both sides and look at prime factors.

28. If a number is a perfect cube, what will its prime factorization look like?

29. Use the fact that $\sqrt{3}$ has already been shown to be irrational.

36. Notice that you do not know which two numbers to add, so you must consider three cases. Write equations to represent each case. You will obtain quadratic equations.

37. Write an equation to express the fact that all 1002 meters of fencing will be used.

38. How would you represent the sum of the rational number and its reciprocal?

39. You might try using variables to represent the lengths of the original two pieces of wire.

Section 9.3

18. When would the cyclist's speed begin to increase?

19. What type of function is $T(n)$? What information do you need to be able to write a formula for a function of this type?

Chapter 10
Section 10.1

23. Use place value in the second equation: $(10a + d)^2 + \ldots = ?$

24. If n^2 has 4 digits, all even, what is the smallest that n can be? What is the largest that n can be?

Section 10.2

17. You must recalculate the class average for 102 students.

19. (a) - (b) You must determine what percentage of scores lie to the left of $z = 1$, $z = 2$, etc., in a normal distribution.

23. (a) How much of the third rectangle together with the first two rectangles on the left will be needed to equal exactly

half the total area of all the
rectangles?

(b) Here you do not have a list of
all the data, but you are given
relative frequencies instead.
Suppose, for example, that
there were exactly 10 scores.
Then how many scores of 10,
20, 30, 40, and 50 would
there be?

24. How much time did it take the dirt
biker to complete his first lap? At
an average speed of 60 mph, how
much time does he have to circle
the track twice?

Chapter 11
Section 11.1

20 Let S = sample space and
A = event the dart hits the bull's
eye. Find $m(S)$ and $m(A)$.

Section 11.2

12. In how many ways can the first
digit be chosen? The second
digit?

20. (d) Remember there is more than
one way to get a matched pair.

22. If one puppy is known to be female,
how does the sample space
change?

23. When a ball is picked at random
from the second box, there are
eleven balls in the box. Notice that
there are two ways to choose a
white ball from the second box.

Section 11.3

26. Remember that the expected
number of trials can be calculated
as $E = v_1p_1 + v_2p_2 + ... + v_np_n$.
In this problem v_i = number of
trials required to get all five prizes
and p_i = probability of needing
that number of trials.

Chapter 12
Section 12.1

14. Remember that a rhombus is a
parallelogram and a square is a
rectangle.

15. Remember that a square is a
rhombus, a rectangle, and a
parallelogram. Also, remember
that a rectangle is a parallelogram.

16. (b) Remember that a kite is not
necessarily a rhombus.

Section 12.2

8. Remember that no two tetrominos
should match - by turning, flipping
over, etc.

Section 12.3

8. Remember that $m(\angle AFE) = 180°$.

10. Remember that the sum of the
angle measures in a triangle is
180°.

11. Use vertical angles to find another
angle that is congruent to $\angle 1$ or
$\angle 2$.

14. What must you show to verify that
quadrilateral ABCD is a
parallelogram?

17. As each new line is added, how many lines does it intersect? How many regions does it divide?

Section 12.4

16. Make a table and look for a pattern in the relationship between the number of people and the number of handshakes.

17. Notice that the angle with measure a is one angle of an isosceles triangle.

18. What is the measure of the base angles in each triangle forming an "arm" of the star?

Chapter 13
Section 13.1

29. What is the standard for length in the English system? In terms of convertibility, consider the relationships between feet and inches, miles and feet, etc. Are similar ratios used?

31. Use dimensional analysis or use the formula $D = R \times T$.

34. (a) You can use dimensional analysis and perform all of the needed conversions at once. Start with the 48 ft^3 of wood cut per hour.
    (d) You can also solve this problem by using dimensional analysis.

35. Remember that you cannot average the speeds (2 kph and 6 kph) to find the average speed. You must divide the total distance traveled by the total time elapsed.

Section 13.2

26. Make a table listing possible values of x and y and the areas that would result.

28. What is the area of one of the right triangles? What is the area of the small square?

40. How could the builders have checked to see that an angle measured 90° by measuring off the sides of a right triangle?

43. Divide the quadrilateral into two triangles. To apply Hero's formula you will need to know the lengths of all three sides of each triangle.

46. Use the radii shown in your drawing to find the difference in the areas of the two circles.

47. If the sum of the areas of all four semicircles is calculated, what areas have been counted twice?

48. What fraction of the area inside the square is the area inside the circle(s)? What percent is that?

49. Find the area of the hexagon by dividing it into six congruent equilateral triangles.

50. Using the radius of the earth as given (6380 km) and the Pythagorean Theorem, you can find the distance from Portland to the north-south axis of the earth.

Section 13.3

10. When you calculate the surface area, remember to count only the

faces of the cubes that are on the outside of the prism.

12. Write equations using the lengths of the edges and the areas of the sides. You should have three equations and three unknowns.

13. Notice that both formulas involve the radius, not the diameter, of a sphere.

16. How will the height of the cylinder be related to the diameter of the sphere?

17. How do the areas of the three stripes compare when the paper is laid flat? It might be helpful to cut out the stripes of each color and form geometric shapes with them in order to compare the areas.

Section 13.4

10. Be careful with units.

15. Be careful with units.

17. What is the volume of the metal in the sphere?

18 (a) Be careful with units.

23. Write equations using the lengths of the edges and areas of the sides of the box. You should have three equations in three unknowns.

Chapter 14
Section 14.1

17. (a) If D, B, and T are collinear, what pair of congruent angles is formed?

(c) The width of the river is AT. What side in $\triangle CBD$ corresponds to side $\overline{AT}$ in $\triangle ABT$?

23. Since angles D and E are supplementary, what equation can be written in terms of x?

Section 14.2

9. What two triangles are similar in your picture?

10. Be sure that you find the distance from the store, not the distance from the edge of the shadow.

14. Notice that both triangles are right triangles. So, given two sides of one of the triangles, the third side can be found.

16. (a) What is the diameter of the surface of the water?
    (b) At any depth, how is the radius of the surface of the water related to the depth of the water?

17. Use similar triangles and a proportion to find AP and PB. Then use another variable to determine CP and PD. For example, let CP = y. What can be used to represent PD?

18. Try using two variables. If b = base and h = height of the smaller triangle, what expressions could be used to represent the base and height of the larger triangle? Calculate the areas of the two triangles.

20. (d) At each step in the process, a segment is broken into four

shorter segments. How is the length of each of the shorter segments related to n?

21. (d) At each step in the process, four smaller triangles are added on to each section of the figure. How is the area of one of these small triangles related to n?

22. Corresponding sides of similar triangles are proportional.

Section 14.3

15. (a) Which segments must be congruent because of the radii used during your construction?
    (b) What pairs of angles are congruent because of the fact that $\triangle APQ \cong \triangle BPQ$, as found in part (a)?
    (c) What can be said about $\angle APR$ and $\angle BPR$ as a result of (b)?
    (d) What can be said about $\overline{AR}$ and $\overline{BR}$ as a result of part (b)?

16. Remember that if two angles of a triangle are congruent to two corresponding angles of another triangle, then the third pair of corresponding angles is also congruent.

Section 14.4

15. What must be true about segments formed on any other transversal?

16. Based on your measurements, do any segments or angles appear to be congruent?

17. What is the product of a and its reciprocal, $1/a$? Use that product

in place of the product ab in the construction. Consider two separate cases: (1) $a > 1$ and (2) $a < 1$.

18. (b) What triangles in the figure are similar? You might find it helpful to look carefully at the proportion $\dfrac{a}{x} = \dfrac{x}{b}$, which is what you want to show. What triangles might lead to this proportion?

Section 14.5

4. What is the definition of a parallelogram?

6. Use the alternate interior angles property to show that pairs of sides are parallel.

7. What can you say about consecutive angles in quadrilateral PQRS?

8. Remember that opposite sides of a parallelogram are congruent.

9. Example 14.13, in this section, may be useful here.

10. You will probably want to use the result proved in problem 4.

13. What is the length of the shorter leg of the right triangle at each end of the trapezoid?

14. You might want to review the similarity properties of triangles. Which one can be applied here?

16. (c) To show $a^2 + b^2 = c^2$, write $a^2 + b^2$ in terms of x and y, substituting the given expressions. Use algebra to

simplify the resulting expression.

17. Show that both ΔB'A'C' and ΔBAC are congruent to ΔBAD, which you constructed, by using the SAS congruence property.

18. If ΔABC ~ ΔA'B'D', what angles must be congruent? Why does this create a contradiction?

Chapter 15
Section 15.1

21. Sketch the three-dimensional shape on graph paper and use the coordinates given to determine its dimensions.

28. If P is the assumed point of intersection of the two lines and *m* is the slope of each of the two lines, what will be the slope of the line segment between P and any other point on either line?

Section 15.2

34. Represent the lengths of the line segments forming the rectangle in terms of *x, y, a,* and *b*. Use the Pythagorean Theorem to write three equations in *x, y, a,* and *b*. Solve simultaneous equations.

Section 15.3

9. (a) Find the equation of the line through two points: A and the midpoint of $\overline{CB}$.

11. (b) Use your result from part (a).

12. You must find the equation of the line that contains point R and that is perpendicular to $\overline{ST}$.

13. What are the midpoints of the diagonals?

15. If MNOP is a parallelogram, then opposite sides must be parallel. How can you show that?

19. Use the distance formula or Pythagorean Theorem to determine the lengths of the non-horizontal sides.

24. Determine the number of possible paths of the same length from A to each point in the grid, starting with the upper right corner. Label each point with the number of paths to that point and look for a pattern.

25. To count the "tilted squares," you might consider the different slopes that are possible for one side of the square.

26. You might try using variables. For example, let *x* = number of rows of black circles and *y* = number of columns of black circles. What expression could be used to represent the number of white circles?

Chapter 16
Section 16.1

38. More than one answer is possible for parts (b) and (d).

39 (a) You might find your result from problem 36(c) helpful here.
    (c) How does the line of reflection relate to each point and its image?
    (d) Think about your answer to part (c).

40. A Mira might be helpful here. Another approach might be to see if you can find some points that are their own images.

41. You might find the transformations given in problem 40 helpful as examples.

Section 16.2

19. To show that two lines are parallel, you could show that corresponding angles are congruent.

21. You might find the argument you used for problem 19 helpful here.

23. (b) How do you locate the center of a rotation? Example 16.8 may be useful here.
     (d) How do you locate the reflection line of a glide reflection? Example 16.8 might be helpful here.

25. You need to find R' such that PQ/PQ' = PR/PR'. Use similar triangles here.

26. $\overline{PR}$ will be parallel to $\overline{P'R'}$ in your construction.

27. If $\overline{PP'}$ intersects l in R and $\overline{QQ'}$ intersects l in point S, you might try drawing segments $\overline{RQ}$ and $\overline{RQ'}$.

Section 16.3

11. Remember that a half-turn is an isometry and so has all the properties of an isometry.

14. (b) Consider the image of each vertex under the reflection M_{AC}.

16. (a) Remember that point A represents any point.
     (b) What is the distance between r and s in terms of x and y ?

17. Remember that the angle of incidence will equal the angle of reflection.

18. What isometry or combination of isometries will map A'B'C'D' to EFGH? Check the orientations of the two squares.

19. Rotate the cube so that $\overline{AB}$ is on the bottom in place of $\overline{DC}$. What is the image of X under this rotation?

PART 2 - SOLUTIONS TO PART A PROBLEMS

from

Mathematics for Elementary Teachers

A CONTEMPORARY APPROACH

by Gary L. Musser and William F. Burger

prepared by Roger J. Maurer and Vikki R. Maurer

SOLUTIONS - PART A PROBLEMS

Chapter 1: Introduction to Problem Solving

Section 1.1

1. Draw a picture like the one shown at left. Notice that there are four small triangles formed by the intersection of the diagonals. Upon further inspection, we see that any two adjacent triangles form another larger triangle. Therefore, there are eight triangles formed.

Section 1.1

2. Let w = the width of the tennis court. Since the length of the court is 6 feet more than twice the width, the length of the court, l, is $2w + 6$. Since the perimeter of a rectangle is $2w + 2l$ and the perimeter of the rectangular tennis court is 228 feet, we know

$$2w + 2l = 228$$
$$2w + 2(2w + 6) = 228 \quad \text{since } l = 2w + 6$$
$$2w + 4w + 12 = 228$$
$$6w + 12 = 228$$
$$6w = 216$$
$$w = 36 \text{ feet}$$
$$l = 2(36) + 6 = 78 \text{ feet}$$

Therefore, the dimensions of the tennis court are 36 feet by 78 feet.

Section 1.1

3. If the number is a multiple of 11 and there are only two digits, a pair, then the two digits must be the same. Our options are 11, 22, 33, 44, 55, 66, 77, 88, and 99. The number is also even so our options become 22, 44, 66, and 88. When we multiply the pair, we automatically have a perfect square so we need only look for the square that is also a cube.

$$2^2 = 4 \qquad \text{not a cube}$$
$$4^2 = 16 \qquad \text{not a cube}$$
$$6^2 = 36 \qquad \text{not a cube}$$
$$8^2 = 64 = 4^3$$

The number must be 88.

Section 1.1

4. The number 9 can be expressed as 4 + 5. Since any even number can be represented as $2n$ for some n, then any odd number can be represented as $2n + 1$ for some n. Further, $2n + 1$ can be written as $(n + n) + 1$ or $n + (n + 1)$, which represents two consecutive numbers.

Section 1.1

5. Since we can use a symbol more than once, we need not use all the symbols. Use the guess and test strategy and remember the order of operations. The solution can be written 6/6 + 6 + 6 = 13, 6 + 6/6 + 6 = 13, or 6 + 6 + 6/6 = 13.

Section 1.1

6. Let a = the number in the top circle, b = the number in the left circle, and c = the number in the right circle. Since the number in each square is the sum of the numbers in the circles on each side of it, we have the following equations:

$$a + b = 41$$
$$a + c = 49$$
$$b + c = 36$$

Since $a + b = 41$, $b = 41 - a$.
Since $a + c = 49$, $c = 49 - a$.
Substituting these into the equation $b + c = 36$ yields

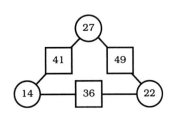

$$41 - a + 49 - a = 36$$
$$90 - 2a = 36$$
$$54 = 2a$$
$$27 = a$$

So $b = 41 - 27 = 14$ and $c = 49 - 27 = 22$.

Section 1.1

7. Systematically draw all possible paths she could take from point A to point B.

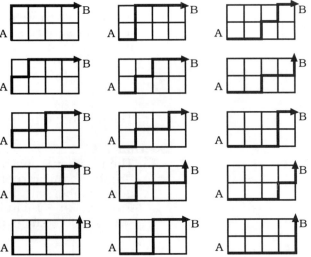

There are 15 paths from A to B.

Section 1.1

8. In order to obtain exactly one liter of water, we must pour water from one pail to the other. Notice two 3-liter pails equal 6 liters, which is one more than a 5-liter pail. Fill the 3-liter pail. Pour it into the 5-liter pail. Fill the 3-liter pail again. Pour as much as you can into the 5-liter pail. This should leave one liter of water in the 3-liter pail.

Section 1.1

9. Draw a picture and use guess and test. Your first instinct may be to place a stool in each of the four corners. In that case, the six remaining stools cannot be placed along the four walls so that each wall has the same number of stools. Continue guessing by placing stools in fewer corners. Also, notice that ten stools divided by four walls yields two stools per wall with two stools left over. By placing these two leftover stools in opposite corners, each wall has three stools.

Section 1.1

10. When we draw a picture and try the guess and test strategy, we notice that there is a systematic way of placing dollars into pockets. If we have zero dollars in the first pocket, then we must put at least one dollar in each remaining pocket. Place one dollar in the second pocket. This forces us to place at least two dollars in each remaining pocket. Place two dollars in the third pocket. Continue in this manner adding one more dollar to each pocket.

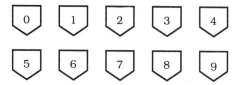

The total number of dollars needed is $0 + 1 + 2 + 3 + 4 + 5 + 6 + 7 + 8 + 9 = 45$. We have only $44. It cannot be done.

Section 1.1

11. We should use a systematic guess and test strategy. Since each corner affects two sides, we should try to determine the corners first.

(i) Try small numbers in the corners as shown at left. Keeping in mind that each side must add to 23, we see that even if we use our largest remaining numbers, 9 and 8, we cannot obtain a side which totals 23.

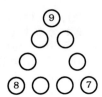

(ii) Try large numbers in the corners as shown. For the side containing 8 and 9, we need 23 − 8 − 9 = 6. Our options are 5 and 1 or 2 and 4. For the side containing 8 and 7, we need 23 − 8 − 7 = 8. Our options are 6 and 2 or 5 and 3. For the side containing 9 and 7, we need 23 − 9 − 7 = 7. Our options are 1 and 6, or 2 and 5, or 3 and 4. The only way to not repeat numbers is to use the combinations 5 and 1, 2 and 6, 3 and 4, respectively.

Section 1.1

12. Looking at the diagram we notice the circle in the upper left hand corner has no arrow leaving it. This circle must contain the 9 since it is a dead end. Notice the circle in the lower right hand corner only has arrows leaving it. This circle must contain 1 since it is a beginning point. Furthermore, you may notice the circle directly below the 9 has only one arrow leaving it leading to the 9. This circle must contain the 8. Now guess and test. Only two circles could contain the number 2. Placing it above the 1 eventually forces us to cross the middle circle twice. Placing the 2 in the middle circle leads to the following solution.

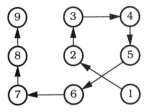

Section 1.1

13. Begin by trying any eight consecutive odd numbers, such as 7, 9, 11, 13, 15, 17, 19, and 21. Their product is 916,620,705 which is much too big. Rather than randomly guessing another set of numbers, eliminate the largest, 21, and include 5. Their product is 218,243,025. This is still too big. When we eliminate the 19 and include 3 the product is the number we were looking for. Therefore, the eight consecutive odd numbers are 3, 5, 7, 9, 11, 13, 15, and 17.

Section 1.1

14. Let x and y represent the two original whole numbers. Since Scott subtracted the two numbers and got 10, we know $x - y = 10$. Since Greg multiplied the two numbers and got 651, we know $xy = 651$. We need to solve the equations $x - y = 10$ and $xy = 651$ simultaneously.

$x - y = 10$ Solve for x.

$x = 10 + y$ Now substitute this into the other equation for x.

$(10 + y)y = 651$

$10y + y^2 = 651$

$y^2 + 10y - 651 = 0$ Factor

$(y - 21)(y + 31) = 0$

$y = 21$ or $y = -31$

Since y is a whole number, we have $y = 21$. Therefore, $x = 10 + 21 = 31$. The correct sum is $21 + 31 = 52$.

Section 1.1

15. Notice that Gary and Bill sit together. Mike and Tom sit together. Since there are only five seats, Howard must sit either between the two pairs or on an end chair.

$$\underline{\text{H}} \quad \_ \quad \underline{\text{H}} \quad \_ \quad \underline{\text{H}}$$

Therefore, Howard must be in one of three seats. Since Bill must sit in the third seat from Howard, Howard cannot sit in the middle chair.

$$\underline{\text{H}} \quad \_ \quad \_ \quad \_ \quad \underline{\text{H}}$$

Therefore, Howard must be in one of two seats. Place Howard in the first seat. Bill must be in the 4th seat. Gary must sit in the 5th seat since placing him in the 3rd seat would split up Mike and Tom. Since Mike must sit in the 3rd seat from Gary, he should be in the 2nd seat. This forces Tom to sit in the middle.

$$\underline{\text{H}} \quad \underline{\text{M}} \quad \underline{\text{T}} \quad \underline{\text{B}} \quad \underline{\text{G}}$$

We see that Bill sat on the other side of Tom.

Section 1.1

16. (a) To maximize the sum, put the largest digits in the greatest place value positions. Put the 9 and 8 in the hundreds places. It follows that the next largest digits should be put in the tens positions. Put the 7 and 6 in the tens places. The 5 and 4 are forced into the ones places. One possible solution is

$$\begin{array}{r} 975 \\ + \ 864 \\ \hline 1839 \end{array}$$

There are 8 possible solutions. These are found by switching digits of the same place value such as the 9 and the 8.

(b) We want the hundreds digits to differ by 1 to minimize the difference. This difference can be reduced further by arranging the remaining digits so that we are forced to borrow as much as possible. Consider the tens and ones places. We want to subtract the largest possible number, 98, from the smallest possible number, 45. This leaves the 7 and 6 to the hundreds places. Therefore, the smallest possible difference is $745 - 698 = 47$.

Section 1.1

17. Because the two inner squares touch more squares than any others, we should place in them the numbers which contact the fewest numbers, i.e., the most extreme: 1 and 8.

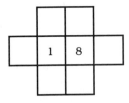

The 1-square touches every other square except the one to the right of the 8. Place the 2 in that square. Similarly, place the 7 in the square to the left of the 1.

| | | 7 | 1 | 8 | 2 | |

Out of the remaining digits, we must place the 3 and 5 together and the 4 and 6 together. Note the 3 cannot touch the 2-square, and the 6 cannot touch the 7-square. We end up with two solutions:

	3	5	
7	1	8	2
	4	6	

or

	4	6	
7	1	8	2
	3	5	

Section 1.1

18. Consider the following figure. Each square has been labeled with a letter.

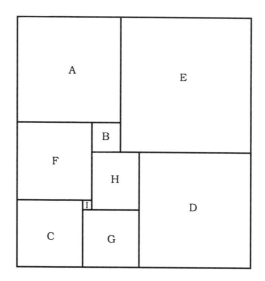

Square	Side Length	Reason
A	14	Given
B	4	Given
C	9	Given
D	15	Given
E	18	Side of A + Side of B = Side of E
F	10	Side of A − Side of B = Side of F
G	8	Side of A + Side of E = 32
		Side of G = 32 − Side of C − Side of D
H	7	Side of H = Side of D − Side of G
I	1	Side of I = Side of G − Side of H

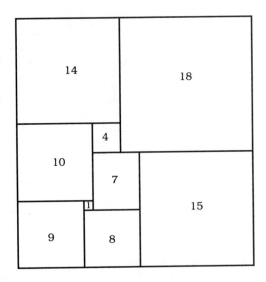

Section 1.1

19. Define two variables. Let s = my son's current age and m = my current age. Since seven years ago my son was one-third my age at the time, $s - 7 = \frac{1}{3}(m - 7)$. Since seven years from now he will be one-half my age at that time, $s + 7 = \frac{1}{2}(m + 7)$. We find my son's age by solving these equations:

$$s - 7 = \tfrac{1}{3}(m - 7) \qquad\qquad s + 7 = \tfrac{1}{2}(m + 7)$$
$$3s - 21 = m - 7 \qquad\qquad\quad 2s + 14 = m + 7$$
$$3s - 14 = m \qquad\qquad\qquad\; 2s + 7 = m$$
$$3s - 14 = 2s + 7$$
$$s = 21$$

My son is 21 years old.

Section 1.1

20. From the problem, we know the year of his birth $= 29 \times$ (his age at death), and we know he had to be born before 1949. Make a list and eliminate unreasonable solutions.

Age at Death	× 29	= Year of Birth	Age in 1949	
64	29	1856	93	Died in 1920
65	29	1885	64	
66	29	1914	35	
67	29	1943	6	Will die in 2010

In 1949, the man could have been 64 or 35 years old.

Section 1.1

21. Notice the P in the sum is alone in the ten thousands place so we can consider it first. P = 1 since it must be the result of carrying. U = 9 since we could not have carried

to obtain P = 1 if U was less than 9. E = 0 since E results from U plus a carry. Notice R + A = 10 and C is odd, since S + S is even and S + S + 1 is odd. So far we have used 0, 1, and 9. Guess and test to solve the rest. R and A could be 2 and 8, 3 and 7, or 4 and 6. If we try R = 8 and A = 2, we see S must be 3 and C must be 7. So we have the following:

$$\begin{array}{r} 9338 \\ + \ 932 \\ \hline 10270 \end{array}$$

Section 1.1

22. Determine the amount of wages lost during the strike. In 22 days, if Kathy worked 6-hour days at $9.74 per hour, then she lost

$$\left(\frac{9.74 \text{ dollars}}{1 \text{ hour}}\right)\left(\frac{6 \text{ hours}}{1 \text{ day}}\right)(22 \text{ days}) = \$1285.68.$$

In the new contract, Kathy must make up $1285.68 in 240 days, working just six hours per day. Let x = the amount of increase in her hourly wage. Since she must make $1285.68 from the increase in her pay during the regular work year, $1440x = 1285.68$, and $x \approx 0.8928$ dollars. Therefore she must receive at least 90¢ more per hour to make up for the lost wages.

Section 1.1

23. Draw a grid and use guess and test. By making 14 moves, the penny and the dime will be exchanged. Consider the original configuration to the left. Notice that each nickel is numbered so that moves can be identified. The moves are shown next.

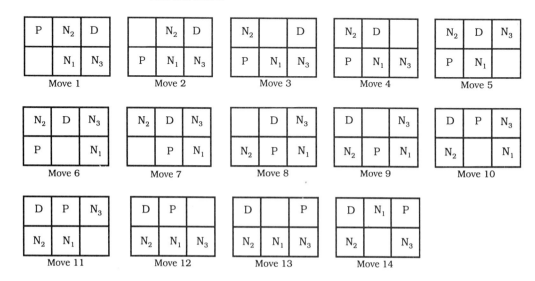

Section 1.1

24. Begin the timers at the same time. Notice we cannot measure the 6 minutes yet since there is only a 3-minute difference in when the timers will go off. As soon as each timer goes off, we restart it. When the 5-minute timer rings for the second time, there will be exactly 6 minutes left on the other timer. Begin measuring the 6 minutes when the 5-minute timer rings for the second time and end when the 8-minute timer rings for the second time.

8-minute timer – – – – – – – –|– – – – – – – –|

5-minute timer – – – – –|– – – – –| (6 min) |
                                begin   end

Section 1.1

25. Let s = the weight of one spool, t = the weight of one thimble, and b = the weight of one button. Since two spools and one thimble balance eight buttons, $2s + t = 8b$. Since one spool balances one thimble and one button, we have $s = t + b$. Since we want an equation relating buttons and spools, we can isolate t in each of the equations above and set them equal to each other.

$$t = 8b - 2s \quad \text{and} \quad t = s - b$$

So $8b - 2s = s - b$
$$9b = 3s$$
$$3b = s$$

So, one spool will balance three buttons.

Section 1.1

26. We do not want consecutive digits next to each other. The numbers in consecutive circles must differ by at least 2. Using the guess and test approach, consider the top four circles. Place in them the odd digits since they differ by multiples of 2.

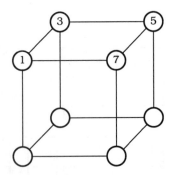

Now we can place the even numbers in order clockwise in the bottom circles as long as the 2 is not directly below the 1 or the 3. A solution is:

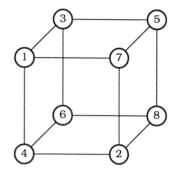

Many solutions are possible. Another solution involves placing the numbers 4, 8, 5, and 1 in the top circles. The numbers 2, 6, 3, and 7 are then placed clockwise in the bottom circles making sure that the 2 is directly below the 4.

Section 1.1

27. Use a variable. Let n be the original number. Perform the operations.

   (1) Add 10: $n + 10$
   (2) Multiply by 4: $4(n + 10) = 4n + 40$
   (3) Add 200: $4n + 40 + 200 = 4n + 240$
   (4) Divide by 4: $(4n + 240)/4 = n + 60$
   (5) Subtract the original number: $n + 60 - n = 60$

Therefore, no matter what the original number was, the result will always be 60 after performing the given set of operations.

Section 1.1

28. The digits must be used in decreasing order. Guess and test by beginning with a number close to 100. We can then add or subtract smaller digits to close in on 100. Begin with 98. Some combination of additions and subtractions on the remaining digits must yield 2. Since the sum of the digits 1 through 7 is 28 we need to subtract somewhere. Since $15 + 13 = 28$ and $15 - 13 = 2$, we want some of the digits to add to 15 and the remaining digits to add to 13. (These will be subtracted.) Notice that $7 + 5 + 3 = 15$, so one solution is
$$98 + 7 - 6 + 5 - 4 + 3 - 2 - 1 = 100.$$

Also, since $6 + 5 + 4 = 15$, another solution is
$$98 - 7 + 6 + 5 + 4 - 3 - 2 - 1 = 100.$$

Many solutions are possible. Other solutions include
$$98 - 7 + 6 + 5 - 4 + 3 - 2 + 1 = 100, \text{ and}$$
$$9 - 8 + 76 + 6 - 5 + 4 - 3 + 21 = 100.$$

Section 1.1
29. Notice that there is a weight restriction for the elevator and that one person must be in the elevator to operate it. The person who weighs 210 pounds must always ride alone, so that person cannot go first. We send the people who weigh 130 pounds and 160 pounds up first. One of these people must ride the elevator back down. It does not matter who comes down, but suppose it is the 130-pound person. Both the 130-pound person and the 210-pound person cannot ride up at the same time. The 210-pound person goes up next, sending the 160-pound person back down. This allows the 160- and 130-pound people to both ride to the top floor.

Section 1.1
30. Make a list of the perfect cubes that are smaller than 1729. These are $1^3 = 1$, $2^3 = 8$, $3^3 = 27$, $4^3 = 64$, $5^3 = 125$, $6^3 = 216$, $7^3 = 343$, $8^3 = 512$, $9^3 = 729$, $10^3 = 1000$, $11^3 = 1331$, $12^3 = 1728$. Find any two of these cubes whose sum is 1729.

$$1^3 + 12^3 = 1729 \quad \text{and} \quad 9^3 + 10^3 = 1729.$$

So $a = 1$, $b = 12$, $c = 9$, and $d = 10$.

Section 1.2
1. (a) Consider the following table.

Sum	Answer
1	$1 = 1^2$
1 + 3	$4 = 2^2$
1 + 3 + 5	$9 = 3^2$
1 + 3 + 5 + 7	$16 = 4^2$
1 + 3 + 5 + 7 + 9	$25 = 5^2$

   Each sum is the square of the number of consecutive odd terms added together.
   (b) Nine are required:
   $1 + 3 + 5 + 7 + 9 + 11 + 13 + 15 + 17 = 81.$
   (c) Thirteen are required:
   $1 + 3 + 5 + 7 + 9 + 11 + 13 + 15 + 17 + 19 + 21 + 23 + 25 = 169.$
   (d) Twenty-three are required.

Section 1.2
2. Compare successive terms in each sequence. Look for a common difference or ratio.
   (a) Each term is half as large as the previous term. The ratio of consecutive terms is 2:1. Therefore, the missing term is $64 \div 2 = 32$.

(b)   Each term is one third as large as the previous term. The ratio of consecutive terms is 3:1. Therefore, the missing term is

$$\frac{1}{9} \div 3 = \frac{1}{9} \times \frac{1}{3} = \frac{1}{27}.$$

(c)   Consider the difference between successive terms.

Notice that the difference increases by 1 each time. Therefore the difference between 16 and the next term should be 5. The missing term is $16 + 5 = 21$.

(d)   Each term is formed by using the previous term in the following way. In the previous term, find the sum of the last two digits. To this sum add the previous term after the last two digits have been replaced with a single zero. That is,

$$\text{new term} = \left(\begin{array}{c}\text{sum of last two digits}\\\text{of the previous term}\end{array}\right) + \left(\begin{array}{c}\text{previous term with last two digits}\\\text{replaced by a zero}\end{array}\right)$$

For example: $12789 = (6 + 3) + (12780) = 9 + 12780$. Therefore, the missing term is $(8 + 9) + (1270) = 17 + 1270 = 1287$.

Section 1.2

3. (a) Notice the letters are in alphabetical order and increase in number by one in each figure. The next figure should contain 5 D's. Notice also in the original figure, which has 2 A's, a segment is drawn (a single side). In the figure with 3 B's, an angle is drawn (two sides). In the figure with 4 C's, a triangle is drawn (three sides). Therefore, in the figure that will contain 5 D's, a four-sided figure, or a square, should be drawn. The figure that best completes the sequence is shown on the left.

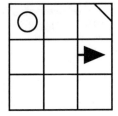

(b) Consider the three types of objects in the figures. In successive figures, the triangle in the upper left-hand square becomes smaller and then changes color. In the last figure given, the triangle has moved to the upper right-hand square. The next figure in the sequence should show a smaller triangle in the upper right-hand square. In successive figures, the arrow moves along the second row of squares, first pointing right, then pointing up in the middle square, then pointing left, and then pointing down in the middle square. The next figure in the sequence should show the arrow pointing right. In successive figures, the circles fill diagonal sets of squares moving from the lower right to upper left. If this pattern continues, the next figure in the sequence should have a single circle in the upper left square. Therefore, the figure that best completes the sequence is shown on the left.

Section 1.2

4. $6^2 - 5^2 = 36 - 25 = 11$
    $56^2 - 45^2 = 3136 - 2025 = 1111$
    $556^2 - 445^2 = 309,136 - 198,025 = 111,111$

(a) Since the number of 5's in the first squared number increases by 1 in each difference, and the number of 4's in the second squared number increases by 1, the next difference should be $5556^2 - 4445^2 = 30,869,136 - 19,758,025 = 11,111,111.$

(b) Make a list and look for a pattern.

Line	No. of 5's in First Squared Number	No. of 4's in Second Squared Number	No. of 1's in Difference
1	0	0	$2 = 2 \times 1$
2	1	1	$4 = 2 \times 2$
3	2	2	$6 = 2 \times 3$
4	3	3	$8 = 2 \times 4$
.	.	.	.
.	.	.	.
.	.	.	.
8	7	7	$16 = 2 \times 8$

Therefore, the eighth line will be
$55555556^2 - 44444445^2 = 1,111,111,111,111,111.$

60	6	10
30	6	5
2	1	2

Section 1.2

5. Notice that each digit in the first column is the product of the two digits directly to the right of it. Also, each digit in the first row is the product of the two digits directly below it. In the third grid:

Consider row 1: $60 = 6 \times ?$, so 10 is the missing number.

Consider column 3: $10 = 5 \times ?$, so 2 is the missing number.

Consider row 3: $2 = ? \times 2$, so 1 is the missing number.

Consider column 2: $6 = ? \times 1$, so 6 is the missing number.

Consider column 1: $60 = ? \times 2$, so 30 is the missing number.

Therefore, the third grid is shown to the left.

Section 1.2

6. (a) Count the dots in each array.

Triangular Number	Number of Dots in Shape
1	$1 = 1$
2	$3 = 1 + 2$
3	$6 = 1 + 2 + 3$
4	$10 = 1 + 2 + 3 + 4$
5	$15 = 1 + 2 + 3 + 4 + 5$
6	$21 = 1 + 2 + 3 + 4 + 5 + 6$

(b)

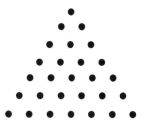

(c) The 10th triangular number will contain $1 + 2 + 3 + 4 + 5 + 6 + 7 + 8 + 9 + 10 = 55$ dots.

(d) The 11th triangular number will have $55 + 11 = 66$ dots. The 12th triangular number will have $66 + 12 = 78$ dots. The 13th triangular number will have $78 + 13 = 91$ dots. Therefore, the triangular number with 91 dots is the 13th number.

(e) The 14th triangular number has $91 + 14 = 105$ dots. The 15th triangular number has $105 + 15 = 120$ dots. The 16th triangular number has $120 + 16 = 136$ dots. The 17th triangular number has $136 + 17 = 153$ dots. Therefore, there is no triangular number that has 150 dots.

(f) Consider the table we made in part (a). Add a third column labeled "Twice the Number of Dots", as shown in the following table.

Triangular Number	Number of Dots in Shape	Twice the Number of Dots
1	1	$2 = 1 \times 2$
2	3	$6 = 2 \times 3$
3	6	$12 = 3 \times 4$
4	10	$20 = 4 \times 5$
5	15	$30 = 5 \times 6$
6	21	$42 = 6 \times 7$
.	.	.
.	.	.
.	.	.
n	?	$n \times (n+1)$

Therefore, the number of dots in each shape is $\dfrac{n(n+1)}{2}$ since it is half as many as twice the number of dots.

(g) The sum of the first 100 counting numbers is $\dfrac{100(101)}{2} = 5050$.

Section 1.2

7. Construct a table comparing the number of dots in each triangular number to the number of dots in each square number.

n	Number of Dots in the nth Triangular Number	Number of Dots in the nth Square Number
1	1	1
2	3	4
3	6	9
4	10	16
5	15	25
6	21	36
7	28	49
8	36	64
9	45	81
10	55	100
.	.	.
.	.	.
.	.	.

(a) The third square number is 9. The two triangular numbers whose sum is 9 are the second and third, which have 3 and 6 dots, respectively.

(b) The fifth square number is 25. The two triangular numbers whose sum is 25 are the fourth and fifth, which have 10 and 15 dots, respectively.

(c) Notice that the sum of the nth and $(n-1)$st triangular numbers is the nth square number. The 10th square number is 100. It is the sum of the 10th and 9th triangular numbers ($55 + 45 = 100$). The 20th square number is $20^2 = 400$. It is the sum of the 20th and 19th triangular numbers ($210 + 190 = 400$). The nth square number is n^2. It is the sum of the nth and $(n-1)$st triangular numbers.

(d) From the table, we notice that 36 is the 8th triangular number and is also a perfect square. Therefore, the 8th triangular number is the 6th square number.

(e) Consider the differences between pairs of square numbers. The following differences yield triangular numbers:

$$
\begin{aligned}
4 - 1 &= 3 \\
25 - 4 &= 21 \\
49 - 4 &= 45 \\
64 - 49 &= 15 \\
64 - 36 &= 28 \\
64 - 9 &= 55 \\
100 - 64 &= 36 \\
169 - 64 &= 105
\end{aligned}
$$

There are others.

Section 1.2

8. On the sheet Linda pulled out, the left half was numbered A4, and the right half was numbered A15. On the back side of the left half, the number must be A3. Therefore, the sheet containing A1 and A2 is still in the paper. By the same reasoning, if A15 is the number on the right half, the other side of the right half must be numbered A16. Since there is one sheet after this one, the two sides would be numbered A17 and A18. Therefore, there are 18 pages in section A of the newspaper.

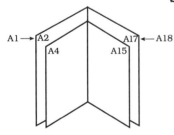

Section 1.2

9. Notice that the sums in parentheses are triangular numbers. From problem 6(f), the sum of the first n counting numbers is $\dfrac{n(n+1)}{2}$. Rewrite each triangular number in parentheses using this formula.

$$1993 \times (1+2+3+4+\ldots+1994) = 1993 \times \frac{(1994)(1995)}{2} = \frac{(1993)(1994)(1995)}{2}$$

$$1994 \times (1+2+3+4+\ldots+1993) = 1994 \times \frac{(1993)(1994)}{2} = \frac{(1993)(1994)(1994)}{2}$$

Since $\dfrac{(1993)(1994)(1995)}{2} > \dfrac{(1993)(1994)(1994)}{2}$, we see that $1993 \times (1+2+3+4+\ldots+1994)$ is larger.

Section 1.2

10. Consider the simpler problem of finding the counterfeit coin given three coins. Begin by weighing two coins. If they weigh the same, the third coin is the counterfeit coin, and we can tell if it is heavier or lighter by comparing it to one of the original coins. If the first two coins do not balance, then switch the heavier one with the third coin. If the third coin is heavier, then the counterfeit coin is the lighter one. If they balance, then the heavy coin is counterfeit.

    Now we return to the problem of five coins. Choose any two coins to weigh. If they balance, use the method from the three-coin problem. If they do not balance, then select one of the remaining three coins and complete the three-coin test.

Section 1.2

11. Make a table of the amount paid each day and the sum of the payments up to that day and look for a pattern.

Day	Pay for Day	Pay to Date
1	1	$1 = 2^1 - 1$
2	2	$3 = 2^2 - 1$
3	4	$7 = 2^3 - 1$
4	8	$15 = 2^4 - 1$
5	16	$31 = 2^5 - 1$

    Notice the pay to date is $2^{\text{day}} - 1$. So on day 30, the total is $2^{30} - 1 = 1{,}073{,}741{,}823$ cents, which is \$10,737,418.23. This amount is much greater than 1 million dollars.

Section 1.2

12. For 1 triangle, the perimeter = 3.
    For 2 triangles, the perimeter = 4.
    For 3 triangles, the perimeter = 5.
    For 4 triangles, the perimeter = 6.

    Notice that the perimeter is always 2 more than the number of triangles. We can summarize this relationship in a table such as the one shown next.

Number of Triangles	Perimeter
1	3
2	4
3	5
4	6
5	7
6	8
.	.
.	.
.	.
10	12
.	.
.	.
.	.
38	40
.	.
.	.
.	.
n	$n + 2$

Section 1.2

13. Since the area is 72 square feet, and the garden has whole-number dimensions, we need to list all possible length × width factor pairs. To find the possible lengths of fence needed, we need to find the perimeter of the garden for each set of dimensions. Recall that Perimeter = 2 × Length + 2 × Width.

Length	Width	Area	Perimeter
1	72	72	146
2	36	72	76
3	24	72	54
4	18	72	44
6	12	72	36
8	9	72	34

Section 1.2

14. (a) Using systematic guess and test, begin by considering numbers with a 1 as the leftmost digit. For example, consider 10, 11, 12, 13, 14, 15, ... Of these, consider only multiples of three since they are the only numbers that could be three times as large as the number with the leftmost 1 removed. Some of the ones that are multiples of 3 are 12, 15, and 18. Of these, only 15 is three times the number with 1 removed (15 = 3 × 5). To find another number with this property, notice that if we add zeros to the numbers under consideration, we still have multiples of three that have 1's as their leftmost digits: 120, 150, and 180. Of these, only 150 is three times the number

with the 1 removed (150 = 3 × 50). We can generate infinitely many of these types of numbers by adding 0's: 1500, 15000, 150000, ...

(b) We know from part (a) that we only need to check numbers that have 1 as the leftmost digit and that are also multiples of 5: 15, 105, 110, 115, 120, 125, 130, 135... Of these, only 125 is 5 times the number with the 1 removed (125 = 5 × 25). Once again, if we add zeros, then we can find other such numbers: 1250, 12500, 125000,...

(c) We need to consider numbers that have 2 as the leftmost digit and that are also multiples of 6: 24, 204, 210, 216... Of these, only 24 is 6 times the number with the 2 removed (24 = 6 × 4). Adding zeros, we can find other such numbers: 240, 2400, ...

(d) We need to consider numbers that have 3 as the leftmost digit and are also multiples of 5: 30, 35, 300, 305, 310, 315, 320, 325... Of these, only 375 is 5 times the number with the 3 removed (375 = 5 × 75). Adding zeros, we can find other such numbers: 3750, 37500...

Section 1.2

15. Make a table to help count zeros. First count zeros that precede nonzero digits.

Numbers	Zeros
0001 to 0009	9 × 3 = 27
0010 to 0099	90 × 2 = 180
0100 to 0999	900 × 1 = 900

Note: This table shows that there are 9 numbers of this type with 3 zeros each, 90 numbers with 2 zeros each, and 900 numbers with 1 zero each. Now count all zeros that follow a nonzero digit.

Numbers	Zeros
0001 to 0099	9 × 1 = 9
0100 to 0999	9 × 2 = 18
	81 × 1 = 81
	9 × 9 = 81
1000	3

Note: This table shows that in the numbers of this type, there are 9 multiples of 10 between 0001 and 0099 with 1 zero each. Of the numbers in the interval 0100 – 0999, there are 9 multiples of 100 with 2 zeros each, 81 multiples of 10 (not multiples of 100) with 1 zero each, and 81 numbers with a zero in the tens digit but no zeros in the ones and hundreds digits. (For example 0101,

0309, 0704, ...) There are 9 for each of the 9 hundreds. Thus $9 \times 9 = 81$. Finally, 1000 contains 3 zeros.

The total number of zeros is the sum of the number of zeros found in the preceding manner:

$27 + 180 + 900 + 9 + 18 + 81 + 81 + 3 = 1299$ zeros.

For each set of one hundred numbers, 0–99, 100–199, 200–299, 300–399, 400–499, 500–599, 600–699, 700–799, and 800–899, the number of nines will be the same. Therefore, we only need to count the number of nines from 0001 to 0100 and multiply by nine. From 0001 to 0100, the numbers requiring a nine are as follows: 0009, 0019, 0029, 0039, 0049, 0059, 0069, 0079, 0089, 0090, 0091, 0092, 0093, 0094, 0095, 0096, 0097, 0098, 0099. We see 20 nines are used. From 0001 to 0899, $9(20) = 180$ nines are used. Finally, from 900 to 999, each number begins with a nine, so in addition to the 20 nines we already know about, there are 100 more nines. Therefore, the total number of nines used is $180 + 120 = 300$.

Section 1.2

16. Continue the number arrangement and look for patterns.

Column 1	Column 2	Column 3	Column 4	Column 5
	2 ---	3	4	5 ---
--- 9	--- 8	$\Big\}+8$ 7	6	$\Big\}+8$
$+8\Big\{$	$+8\Big\{$ 10 ---	11	12	13 ---
--- 17	--- 16	$\Big\}+8$ 15	14	$\Big\}+8$
$+8\Big\{$	$+8\Big\{$ 18 ---	19	20	21 ---
--- 25	--- 24	23	22	

(a) Notice that 100 is even and there are only two even columns, 2 and 4. Consider column 2. In it the difference between every other element is 8. The integers in column 2 are either $2 + 8n$ or $0 + 8n$, for some whole number n. If 100 is in column 2, it would have one of these two forms. Because $100 = 4 + 8 \times 12$, it cannot be in this column. Thus, 100 must be in column 4.

(b) Similarly, 1000 must be in column 2 or 4. Notice that $1000 = 0 + 8 \times 125$, so it must be in column 2.

(c) Because 1999 is odd, we have three columns to consider. Notice column 1 contains integers of the

1999 = 7 + 8 × 249. Thus, it cannot be in column 1 or 5. The number 1999 must fall in column 3.

(d) By the same reasoning, 99,997 must be in column 5 because 99,997 = 5 + 8 × 12499.

Section 1.2

17. Notice there are three sizes of triangles: $1 \times 1 \times 1$, $2 \times 2 \times 2$, and $3 \times 3 \times 3$. Count the number of triangles of each size separately.

There are nine $1 \times 1 \times 1$ triangles.

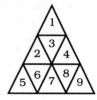

There are three $2 \times 2 \times 2$ triangles.

 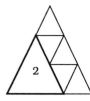

There is one $3 \times 3 \times 3$ triangle.

Thus, there are 9 + 3 + 1 = 13 equilateral triangles of all sizes in the $3 \times 3 \times 3$ triangle.

Section 1.2

18. To find the number of cubes in the 100th collection of cubes, make a list of the number of cubes found in the 1st, 2nd, 3rd, ... collections. Look for a relationship between the dimensions of the solid and the number of $1 \times 1 \times 1$ cubes in the solid. The first solid is a single $1 \times 1 \times 1$ cube. The second solid contains two layers of four cubes for a total of eight $1 \times 1 \times 1$ cubes. The third solid contains three layers of nine cubes each for a total of twenty-seven $1 \times 1 \times 1$ cubes.

Dimensions	Number of Cubes
$1 \times 1 \times 1$	$1 = 1^3$
$2 \times 2 \times 2$	$8 = 2^3$
$3 \times 3 \times 3$	$27 = 3^3$

Notice the number of cubes in each solid is the product of the dimensions. In other words, the number of cubes is the cube of the length of a side . For the 100th collection of cubes, the number of $1 \times 1 \times 1$ cubes is $100(100)(100) = 100^3 = 1,000,000$.

Section 1.2
19. The next six terms of the Fibonacci sequence are 34, 55, 89, 144, 233, and 377. Notice the sum of the squares of the first n Fibonacci numbers is the product of the nth and $n + $ 1st terms of the Fibonacci sequence. To predict the sum of $1^2 + 1^2 + 2^2 + 3^2 + ... + 144^2$, we need to find the Fibonacci number that is one term beyond 144. We predict the desired sum is equal to $144 \times 233 = 33,552$.

Section 1.2
20. Sixteen terms of the Fibonacci sequence are: 1, 2, 3, 5, 8, 13, 21, 34, 55, 89, 144, 233, 377, 610, 987, 1597.
We observe the following pattern:

$$1 + 2 = 3$$
$$1 + 2 + 5 = 8$$
$$1 + 2 + 5 + 13 = 21$$
$$1 + 2 + 5 + 13 + 34 = 55$$
$$1 + 2 + 5 + 13 + 34 + 89 = 144$$

Compare each equation with the Fibonacci sequence. Notice that the total for each set of additions is the element in the Fibonacci sequence that is one term beyond the largest Fibonacci number in the sum. Thus, the answer to $1 + 2 + 5 + 13 + 34 + 89 + 233 + 610$ should be 987. Check this using your calculator.

Section 1.2
21. (a) Consider the following table.

Numbers in the Diagonals of Pascal's Triangle	Sum
1	1
1	1
11	2
12	3
131	5
143	8
1561	13

(b) Notice that the sequence of numbers in the "Sum" column is the Fibonacci sequence. Therefore, the next three sums should be 21, 34, and 55. If we add entries to Pascal's triangle, we have:

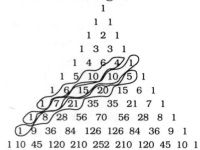

Numbers in Diagonal	Sum
1 6 10 4	21
1 7 15 10 1	34
1 8 21 20 5	55

Section 1.2

22. (a) The encircled numbers are 1, 2, 1, 3, 3, 4, and 6. Their sum is 20.

(b) Consider the following encircled numbers in Pascal's Triangle:

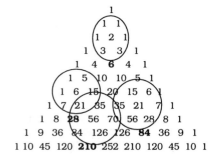

Encircled Numbers	Sum
1, 1, 1, 2, 1, 3, 3	12 = 2(6)
1, 5, 1, 6, 15, 7, 21	56 = 2(28)
15, 6, 35, 21, 7, 56, 28	168 = 2(84)
21, 35, 28, 56, 70, 84, 126	420 = 2(210)

Notice the sum of each set of seven numbers is twice the number that is directly below the center term in each circle.

Section 1.2
23. Determine the time for a round trip for each of the two days. Recall that Distance = Rate × Time, so

$$\text{Time} = \frac{\text{Distance}}{\text{Rate}}.$$

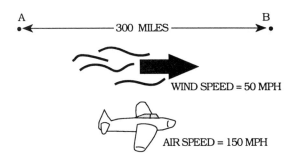

A ←————— 300 MILES ————→ B

WIND SPEED = 50 MPH

AIR SPEED = 150 MPH

Today: The sum of the wind speed and the air speed is the ground speed of the plane. Therefore, the ground speed = 50 mph + 150 mph = 200 mph. In traveling from A to B, he will travel 300 miles at 200 mph so the time required is $\frac{300}{200}$ = 1.5 hours. On the return trip, his ground speed will be 150 mph – 50 mph = 100 mph. Therefore, in traveling from B to A, he will travel 300 miles at 100 mph. The time required will be $\frac{300}{100}$ = 3 hours. The total time required for the round trip this morning will be 1.5 + 3 = 4.5 hours.

Tomorrow: If he waits until tomorrow when there is no wind, he will travel a total distance of 600 miles at 150 mph, so the time required will be $\frac{600}{150}$ = 4 hours.

Therefore, his travel time will be shorter if he waits until tomorrow.

Section 1.2
24. (a)

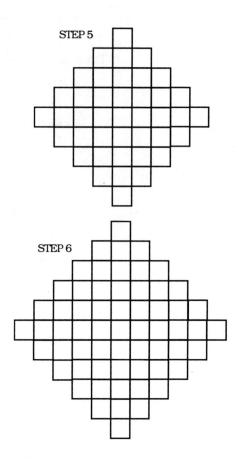

STEP 5

STEP 6

(b) Consider the following table.

Step Number	Number of Squares	Number of New Squares Attached
1	1	
2	5 = 1 + 4	4 = 4(1)
3	13 = 1 + 4 + 8	8 = 4(2)
4	25 = 1 + 4 + 8 + 12	12 = 4(3)
5	41 = 1 + 4 + 8 + 12 + 16	16 = 4(4)
6	61 = 1 + 4 + 8 + 12 + 16 + 20	20 = 4(5)

Notice that the number of squares attached at each step is a multiple of 4. Each figure is formed by adding to the original square. Consider the sum for each step. If we factor out a 4 from each term, with the exception of the 1, we obtain the triangular numbers. This factorization is illustrated in the next table.

Step Number	Number of Squares
1	1
2	$5 = 1 + 4 = 1 + 4(1)$
3	$13 = 1 + 4 + 8 = 1 + 4(1 + 2)$
4	$25 = 1 + 4 + 8 + 12 = 1 + 4(1 + 2 + 3)$
5	$41 = 1 + 4 + 8 + 12 + 16 = 1 + 4(1 + 2 + 3 + 4)$
.	.
.	.
.	.
n	$1 + 4(1 + 2 + 3 + ... + (n - 1))$

Since the formula for the sum of the first $n - 1$ counting numbers is $\dfrac{(n-1)(n)}{2}$, the number of squares for the nth step is

$$1 + 4(1 + 2 + 3 + ... + (n - 1)) = 1 + \frac{4(n-1)(n)}{2} = 1 + 2n(n-1).$$

(c)   At step 7, the sum is $1 + 2(7)(7-1) = 1 + 2(7)(6) = 85$.

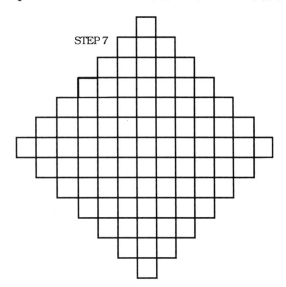

STEP 7

(d)   In the 10th figure, there would be $1 + 2(10)(9) = 1 + 180 = 181$ squares. In the 20th figure, there would be $1 + 2(20)(19) = 1 + 760 = 761$ squares. In the 50th figure, there would be $1 + 2(50)(49) = 1 + 4900 = 4901$ squares.

Section 1.2

25. There is an unlimited supply of darts. Begin by listing integers. Notice which ones can be obtained using some combination of 9's and 4's. Make a table. For each possible score note the number of 9's and 4's used.

Integer	Possible Score?	Number of 9's Used	Number of 4's Used
1	No		
2	No		
3	No		
4	Yes	0	1
5	No		
6	No		
7	No		
8	Yes	0	2
9	Yes	1	0
10	No		
11	No		
12	Yes	0	3
13	Yes	1	1
14	No		
15	No		
16	Yes	0	4
17	Yes	1	2
18	Yes	2	0
19	No		
20	Yes	0	5
21	Yes	1	3
22	Yes	2	1
23	No		
24	Yes	0	6
25	Yes	1	4
26	Yes	2	2
27	Yes	3	0
28	Yes	0	7
29	Yes	1	5
30	Yes	2	3
31	Yes	3	1
32	Yes	0	8
33	Yes	1	6
34	Yes	2	4
35	Yes	3	2
.	.	.	.
.	.	.	.
.	.	.	.

The largest score that cannot be obtained is 23. Notice a pattern in the number of 9's used. Once the score of 27 is reached, each of the following scores is formed by adding zero, one, two, or three 9's and a multiple of 4.

Section 1.2

26. Notice that a new point is added to any flat surface to create a new star.

   (a) For the third star, notice that each of the 6 points of the second star receives 2 new points so that each appears to be made up of 3 small points. Therefore, the third star is made up of $3 \times 6 = 18$ points. One of the six clusters of three points from the third star is shown to the left. Notice that there are 8 flat surfaces.

   (b) To form the fourth star, notice that there are 8 flat surfaces for each of the clusters of 3 points from the third star. Add a new point to each of the flat surfaces. The fourth star will have $11 \times 6 = 66$ points. One of the six clusters of 11 points from the fourth star is shown next.

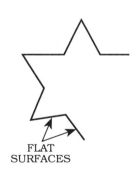

FLAT
SURFACES

SOLUTIONS - PART A PROBLEMS

Chapter 2: Sets, Whole Numbers, Numeration, and Functions

Section 2.1

24. (a) X∩Y: Consider the Venn Diagram.

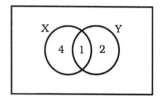

Set X has 5 elements. Set Y has three elements. Set X∩Y has one element. Notice that all of the elements in Y could also be in X, which would maximize the number of elements in the intersection. Therefore, the greatest number of elements possible in X∩Y is 3.

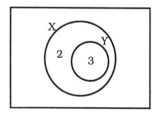

X∪Y: Consider the Venn Diagram.

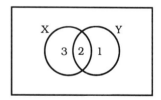

The sum of the number of elements in X∪Y is 6. If we had fewer elements in X∩Y, then there would be more in X∪Y. The greatest number of elements possible in X∪Y is 8.

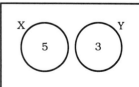

(b) X∩Y:

The number of elements in the intersection of X and Y cannot be larger than the number of elements in either set individually. Recall part (a). Suppose set Y is totally contained in set X. In that case, the maximum number of elements in X∩Y is y, which represents the number of elements in set Y.

From part (a), we see that the greatest number of elements in the union occurs when the sets have no intersection, in which case the number of elements in X∪Y is the sum of the number of elements in the individual sets. The maximum number of elements in X∪Y is $x + y$.

Section 2.1

25. Consider sets A and B. The element 1 in set A can be paired with any of the four elements in set B. After the initial pairing, element 2 in set A could be paired with any of the three remaining elements in set B. Likewise, element 3 in set A could be paired with either of the two remaining elements in set B, and element 4 in set A must be paired with the remaining element in set B. Therefore, the total number of one-to-one correspondences possible from A to B is $4 \times 3 \times 2 \times 1 = 24$.

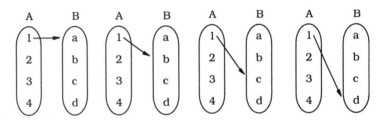

Section 2.1

26. (a) Consider the set { }. Its only subset is itself: { }, so it has 1 subset.

    (b) Consider the set {x}. Its subsets are { } and {x}, so it has 2 subsets.

    (c) Consider the set {x, y}, its subsets are { }, {x}, {y}, and {x, y}, so it has 4 subsets.

    (d) Consider the set {x, y, z}. Its subsets are { }, {x}, {y}, {z}, {x, y}, {x, z}, {y, z}, and {x, y, z}, so it has 8 subsets.

    (e) Construct a table and notice a relationship between the number of elements in a set and the number of subsets.

Number of Elements	Number of Subsets
0	$1 = 2^0$
1	$2 = 2^1$
2	$4 = 2^2$
3	$8 = 2^3$

Notice that the number of subsets is always 2 raised to a power. The exponent is the number of elements in the set. Therefore, for a set of 5 elements, the number of subsets is $2^5 = 32$.

(f) We can generalize to the case of n elements. The number of subsets is 2^n.

Section 2.1

27. (a) This is possible. For example, consider set A = {1, 2, 3} and set B = {x, y, z}. Sets A and B are equivalent since we can find a 1-1 correspondence between them. However, sets A and B cannot be equal since they do not contain exactly the same elements.

(b) This is not possible. Two sets that are equal contain exactly the same elements, thus the same number of elements. We can always find a 1-1 correspondence between sets with the same number of elements.

Section 2.1

28. (a) Using a Venn Diagram one can see that $D \cap E = D$ only when $D \subseteq E$.

(b) Using a Venn Diagram one can see that $D \cup E = D$ only when $E \subseteq D$, or when $E = \varnothing$.

(c) Considering parts (a) and (b), we can see that $D \cap E = D \cup E$ if and only if $D = E$.

Section 2.1

29. Consider a set of eight skirts, S={S1, S2, S3, S4, S5, S6, S7, S8} and a set of seven blouses, B ={B1, B2, B3, B4, B5, B6, B7}. Since any outfit Carme chooses will consist of a single skirt and a single blouse, the concept of Cartesian Product can be applied.

$$S \times B = \begin{cases} (S1, B1) & (S1, B2) & (S1, B3) & \ldots & (S1, B7) \\ (S2, B1) & (S2, B2) & (S2, B3) & \ldots & (S2, B7) \\ \vdots & \vdots & \vdots & \vdots & \vdots \\ \vdots & \vdots & \ldots & \ldots & \vdots \\ (S8, B1) & (S8, B2) & \ldots & \ldots & (S8, B7) \end{cases}$$

The number of outfits Carme can wear is $8 \times 7 = 56$, which is the number of elements in the Cartesian Product.

Section 2.1

30. Consider a simpler problem. Suppose we begin with 8 participants. In the first round, there would be 4 matches. In the second round, there would be 4 participants in 2 matches. In the third round, there would be 2 participants in one match. This yields a total of 4 + 2 + 1 = 7 matches. Alternatively, consider beginning with 5 participants. In round one, 4 of the 5 participants compete in 2 matches, eliminating 2 players. Two of the 3 remaining participants compete in one match, eliminating 1 player. The two remaining participants compete in one final match. This yields a total of 2 + 1 + 1 = 4 matches. Notice that there is a pattern. The number of matches in each case is one less than the number of original participants. Thus, if we begin with 32 participants, the total number of matches is 31.

Section 2.1

31. Draw a Venn diagram. The three sets are T = set of people who watched television, N = set of people who read the newspaper, and R = set of people who listened to the radio. Begin filling the Venn diagram from the inside and work out. Begin with $T \cap N \cap R$ = set of people who watched television *and* read the newspaper *and* listened to the radio. We are given that the number of voters in $T \cap N \cap R$ is 6. Now consider the region $N \cap R$ = set of people who read the newspaper *and* listened to the radio. Notice that $N \cap R$ also contains people who fall in the set $T \cap N \cap R$, so be careful not to count some of the voters twice. Since the set $N \cap R$ contains 9 voters, and we have already accounted for 6 of them, there must be 3 in the remaining portion of $N \cap R$. Similarly, in the remaining portion of $N \cap T$, there would be 20 − 6 = 14 people, and in the remaining portion of $T \cap R$, there would be 27 − 6 = 21 people. Now, 65 people belong in set T. Notice that set T overlaps other sets so we have already accounted for 21 + 6 + 14 = 41 of them. Therefore, there are 65 − 41 = 24 voters in the remaining portion of T. Similarly, there are 39 − (14 + 6 + 3) = 16 in the remaining portion of N, and 39 − (21 + 6 + 3) = 9 in the remaining portion of R. Since there were 100 voters altogether, and we accounted for only 24 + 14 + 16 + 21 + 6 + 3 + 9 = 93 of them, there must be 7 voters who do not rely on any of these methods to keep up with current events.

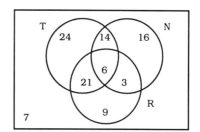

(a) Those voters who kept up with current events by some means other than the three sources belong to the set $\overline{T \cup N \cup R}$. There are 7 voters in $\overline{T \cup N \cup R}$.

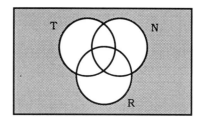

(b) Those voters who kept up with current events by reading the paper but not by watching television belong to the set $N \cap \overline{T}$. There are 19 voters in $N \cap \overline{T}$.

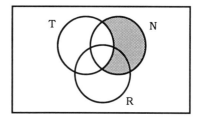

(c) Those voters who use only one of the three sources to keep up with current events belong in the following portions of the Venn diagram.

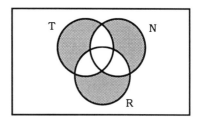

There are 49 voters in this set.

Section 2.1

32. At the convention there are butchers, bakers, and candlestick makers. (Note: Someone could be both a butcher *and* a baker or some other combination.) For each description, consider a Venn Diagram and shade the described region. For example, the fact that 50 persons were in both B and A , but not in set C means that 50 were in the set (A∩B)–C. 50 people fall into that region (the shaded region).

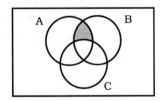

The total number represented in the diagram below is 50 + 70 + 60 + 40 + 50 + 80 = 350. Since there were 375 people at the convention, there must have been 375 – 350 = 25 people who were butchers, bakers, and candlestick makers. Therefore, there were 25 people in A, B, and C.

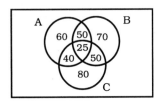

Section 2.1

33. Draw a picture. Construct line segments which connect points on base $\overline{AB}$ to points on the sides. If you use line segments perpendicular to the base, then a 1-1 correspondence can be found.

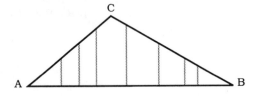

If we continue in the same manner, no points on either the base or the sides will be left out.

Section 2.1

34. Draw a picture. Construct line segments systematically from chord $\overline{AB}$ to arc ACB. A good way to do this is to use line segments that are perpendicular to chord $\overline{AB}$.

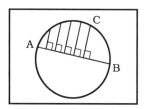

Continuing to draw line segments in this way will yield a 1-1 correspondence between the points on chord $\overline{AB}$ and the points on arc ACB.

Section 2.2

15. The Roman System is a positional (additive and subtractive) system. Confusion can occur anytime there is an additive versus subtractive case, such as IV versus VI, where in the first instance the 1 is subtracted from the 5 and in the second, it is added. There are many such examples: IX versus XI, CM versus MC, and so on. The Egyptian System is not positional. It is a completely additive system. Reversing symbols will not change the value of the number.

Section 2.2

16. (a) Translate each problem to Egyptian symbols and count the symbols used.

    (i)   ∩∩∩∩∩| | | | | | | | + ∩∩∩∩∩∩∩∩| | | | | | | , 30 symbols
    (ii) ?∩∩∩∩∩ - ∩∩∩∩∩∩∩∩∩| | | | | | | | | , 24 symbols
    (iii) ⌇⌇⌇⌇⌇⌇⌇ ????????∩∩∩∩∩∩∩∩| | | | | | |
          + ?????????∩∩∩| | | | , 47 symbols
    (iv) ⌇⌇⌇⌇⌇⌇⌇⌇ ???????∩∩∩∩∩∩∩∩∩| | | | | | | | |
          - ⌇⌇⌇⌇⌇???∩∩∩∩∩∩∩∩| | | | | | | | | , 57 symbols

    (b) Each digit tells us the number of symbols needed of a particular value. The sum of the digits in the number in our system corresponds to the number of symbols needed to represent the number in the Egyptian System. Therefore, to find the number of symbols needed to represent the problem we need to add the digits of the addends (or subtrahend and minuend).

Section 2.2

17. Since the new 1993 Lincoln Mark VIII's are coming out on December XXVI (December 26th), the advertisement probably is being run in a 1992 paper. (It is not likely that the dealer would begin advertising his new 1993 model at the end of 1993). Write 1992 in Roman numerals. Recall that the Roman system has a subtractive principle.

$$\text{M} \underbrace{\text{M C}}_{\text{subtract}} \underbrace{\text{X C}}_{\text{subtract}} \text{I I}$$

M = 1000, CM = 900, XC = 90, and II = 2. Therefore, 1992 = MCMXCII.

Section 2.3

19. Consider a simpler problem. How many digits would be used in a 25-page book? (Note: The front and back of each sheet of paper in a book are separate pages.) Page numbers: 1, 2, 3, 4, 5, 6, 7, 8, 9, 10, 11, 12, 13, 14, 15, 16, 17, 18, 19, 20, 21, 22, 23, 24, 25. Notice that the page numbers from 1 to 9 each have 1 digit. Numbers 10 to 25 each have 2 digits. Total number of digits used in a 25-page book: (9 pages)(1 digit) + (16 pages)(2 digits) = 41 digits in all. Now to solve the given problem, set up a table relating page numbers to digits used.

Page Number	Digits	Total Digits
1 – 9	1	$9 \times 1 = 9$
10 – 99	2	$90 \times 2 = 180$
100 – 999	3	$900 \times 3 = 2700$
1000 – 9999	4	$9000 \times 4 = 36{,}000$

Our book used 2989 digits. By adding the total number of digits on pages numbered from 1 to 999, we see that we have used only 2889 digits. There are 100 digits left to use. Each page from 1000 on uses 4 digits. Since 100/4 = 25 pages, there are 25 pages left. These would be pages 1000 through 1024. Thus, there are 1024 pages in the book.

Section 2.3

20. Use expanded form to represent the two-digit number. Let the two digit number be ab. In expanded form the number would be $10a + 1b$. Now set up two equations using the information given.

(1) The sum of the digits is 12, so $a + b = 12$.
(2) If the digits are reversed, the new number (ba) is 18 greater than the original (ab), so we have $10b + 1a = 10a + 1b + 18$.

Collecting like terms in (2), we have $-9a + 9b = 18$. Now we have two equations and two unknowns. Solving the system of equations $a + b = 12$ and $-9a + 9b = 18$

simultaneously yields $a = 5$ and $b = 7$. Thus, the original two-digit number was 57.

Section 2.3

21. (a) Let's try this sequence of operations using my birthday of 03/17/66.

   (1) Multiply the month by 4: $4(03) = 12$
   (2) Add 13: $12 + 13 = 25$
   (3) Multiply by 25: $25(25) = 625$
   (4) Subtract 200: $625 - 200 = 425$
   (5) Add the day of the month: $425 + 17 = 442$
   (6) Multiply by 2: $442(2) = 884$
   (7) Subtract 40: $884 - 40 = 844$
   (8) Multiply by 50: $844(50) = 42200$
   (9) Add last two digits of birth year: $42200 + 66 = 42266$
   (10) Subtract 10500: $42266 - 10500 = 31766$

   If the final answer is of the form *abcdef*, then the digits in the tens and ones places, *ef*, are the last two digits of the birth year. The digits in the thousands and hundreds places (*cd*) are the digits of the day of the month. The digits in the hundred thousands and ten thousands places (*ab*) are the digits in the month of the year.

   (b) Use expanded notation. If the birthday is *ab/cd/ef*, then the final answer should be
   $$100{,}000a + 10{,}000b + 1000c + 100d + 10e + f$$

(1) Multiply month by 4: $4(10a + b) = 40a + 4b$
(2) Add 13: $40a + 4b + 13$
(3) Multiply by 25: $25(40a + 4b + 13) = 1000a + 100b + 325$
(4) Subtract 200: $1000a + 100b + 325 - 200 = 1000a + 100b + 125$
(5) Add the day: $1000a + 100b + 125 + 10c + d = 1000a + 100b + 10c + d + 125$
(6) Multiply by 2: $2(1000a + 100b + 10c + d + 125) = 2000a + 200b + 20c + 2d + 250$
(7) Subtract 40: $2000a + 200b + 20c + 2d + 250 - 40 = 2000a + 200b + 20c + 2d + 210$
(8) Multiply by 50: $50(2000a + 200b + 20c + 2d + 210) = 100{,}000a + 10{,}000b + 1000c + 100d + 10500$
(9) Add last two digits of year: $100{,}000a + 10{,}000b + 1000c + 100d + 10500 + 10e + f$
(10) Subtract 10500: $100{,}000a + 10{,}000b + 1000c + 100d + 10e + f + 10500 - 10500 = 100{,}000a + 10{,}000b + 1000c + 100d + 10e + f$

   Since the expanded form of *abcdef* is $100{,}000a + 10{,}000b + 1000c + 100d + 10e + f$, we see this technique will always work.

Section 2.3

22. Write out the numbers on both sides of the equation using expanded form. Use a variable for the missing base.

(a) $32 = 44_b$

$3 \times 10 + 2 \times 1 = 4b + 4 \times 1$ Solve for the unknown base.

$30 + 2 - 4 = 4b$

$28 = 4b$

$7 = b$

So $32 = 44_7$

(b) $57_8 = 10_b$

$5 \times 8 + 7 \times 1 = 1 \times b + 0 \times 1$

$40 + 7 = b$

$47 = b$

So $57_8 = 10_{47}$

(c) $31_4 = 11_b$

$3 \times 4 + 1 \times 1 = 1 \times b + 1 \times 1$

$12 + 1 - 1 = b$

$12 = b$

So $31_4 = 11_{12}$

(d) $15_x = 30_y$

$1x + 5 \times 1 = 3y + 0 \times 1$

$x + 5 = 3y$

$x = 3y - 5$

Notice that $x = 3y - 5$ is an equation for a line, so there are an infinite number of solutions. However, the original equation puts restrictions on the solution. Consider base x. Notice that there is a 5 in the ones position. This forces base x to be 6 or greater since base 6 uses only digits zero through five. A lower base would not include the digit five. Therefore, $x = 3y - 5$ and $x > 5$.

Section 2.3

23. List numerals in each base to determine the ones digit.

(a) Base 10: 0, 2, 4, 6, 8, 10, 12, 14, 16, ... Notice the ones digit in each case is either 0, 2, 4, 6, or 8.

(b) Base 4: $0 = 0_4, 2 = 2_4, 4 = 10_4, 6 = 12_4, 8 = 20_4, 10 = 22_4, ...$ Notice the ones digit is always 0 or 2.

(c) Base 2: $0 = 0_2, 2 = 10_2, 4 = 100_2, 6 = 110_2, 8 = 1000_2, 10 = 1010_2, ...$ Notice that the ones digit is always 0.

(d) Base 5: $0 = 0_5, 2 = 2_5, 4 = 4_5, 6 = 11_5, 8 = 13_5, 10 = 20_5, 12 = 22_5, ...$ Notice that the ones digit is always 0, 1, 2, 3, or 4. Base 5 uses only digits 0, 1, 2, 3, and 4 so there can be no others.

Section 2.3

24. Consider a simpler problem such as determining a 3-digit number in base 10. Each of the 3-digits would require at most 9 questions for a maximum total of $3 \times 9 = 27$ questions needed to determine the number. On the other hand, if the 3-digit number was written in a smaller base,

hand, if the 3-digit number was written in a smaller base, say base 4, each digit would require at most 3 questions. Since the largest 3-digit number, 999, is between $4^4 = 256$ and $4^5 = 1024$, the converted number in base 4 would contain at most 5 digits. Therefore, at most $3 \times 5 = 15$ questions would be needed. We see fewer questions are needed when we use a smaller base. In the original problem, we would have the person convert the number to base 2. Since the largest number, 9,999,999, is between $2^{23} = 8,388,608$ and $2^{24} = 16,777,216$, the base 2 number will have at most 24 digits. At most $1 \times 24 = 24$ questions would be needed to determine the number. (The base 10 phone number would require at most $7 \times 9 = 63$ questions.) Once the number is determined, convert back to base 10.

Section 2.4

11. (a) In each case, let n = the number in parentheses.

$$f(0) = \frac{9}{5}(0) + 32 = 32$$

$$f(100) = \frac{9}{5}(100) + 32 = 180 + 32 = 212$$

$$f(50) = \frac{9}{5}(50) + 32 = 90 + 32 = 122$$

$$f(-40) = \frac{9}{5}(-40) + 32 = -72 + 32 = -40$$

(b) $g(32) = \frac{5}{9}(32 - 32) = \frac{5}{9}(0) = 0$

$g(212) = \frac{5}{9}(212 - 32) = \frac{5}{9}(180) = 100$

$g(104) = \frac{5}{9}(104 - 32) = \frac{5}{9}(72) = 40$

$g(-40) = \frac{5}{9}(-40 - 32) = \frac{5}{9}(-72) = -40$

(c) If degrees Celsius = degrees Fahrenheit, then $f(n) = n$ and $g(m) = m$. Notice from part (a) that $f(-40) = -40$ and from part (b) that $g(-40) = -40$. Thus, $-40°C = -40°F$.

Section 2.4

12. Study the sequence of numbers. Notice that there is a common difference of 8 between consecutive terms.

Term	Number
1	21
2	$29 = 21 + 1(8)$
3	$37 = 21 + 2(8)$
4	$45 = 21 + 3(8)$
.	.
.	.
.	.
n	$21 + (n - 1)(8)$

Notice that each term is made up of the sum of 21 and a certain number of 8's. The number of 8's in the sum is always 1 less than the term number. The 458th number is $21 + (458 - 1)(8) = 21 + 457(8) = 3677$.

Section 2.4

13. (a) Since there is a \$35 per month charge, after x months, it would cost $35x$ dollars. However, there is also an initiation fee of \$85. Therefore, the total cost for x months is $C(x) = 35x + 85$.

(b) $C(18) = 35(18) + 85 = 630 + 85 = 715$. Thus, the total amount spent after 18 months by a member was \$715.

(c) We want to find x when $C(x) > 1000$. If $C(x) > 1000$, then $35x + 85 > 1000$. We must solve for x, as shown next.

$$35x > 1000 - 85$$
$$35x > 915$$
$$x > 26.14$$

Since $C(26) = 35(26) + 85 = 995$ and $C(27) = 35(27) + 85 = 1030$, we see at 27 months the cost first exceeds \$1000.

Section 2.4

14. A geometric sequence is of the form ar^{n-1} where r is the common ratio.

(a) Since the second term is 1200, we know
$$1200 = ar^{2-1} = ar .$$
Since the fifth term is 150, we know
$$150 = ar^{5-1} = ar^4 .$$
Now we have two equations and two unknowns.

$1200 = ar$ \qquad Solve for a

$\dfrac{1200}{r} = a$ \qquad Substitute into $150 = ar^4$

$$150 = \frac{(1200)}{r} r^4$$

$$150 = 1200r^3$$

$$\frac{150}{1200} = r^3$$

$$\frac{1}{8} = r^3$$

$$\frac{1}{2} = r$$

(b) To find the first six terms of the sequence, we must determine a. Use $r = \dfrac{1}{2}$ from part (a).

Since $1200 = a\left(\frac{1}{2}\right)$, we see $2400 = a$.

Term	Number
1	$2400\left(\dfrac{1}{2}\right)^{1-1} = 2400\left(\dfrac{1}{2}\right)^{0} = 2400$
2	$2400\left(\dfrac{1}{2}\right)^{2-1} = 2400\left(\dfrac{1}{2}\right)^{1} = 1200$
3	$2400\left(\dfrac{1}{2}\right)^{3-1} = 2400\left(\dfrac{1}{2}\right)^{2} = 2400\left(\dfrac{1}{4}\right) = 600$
4	$2400\left(\dfrac{1}{2}\right)^{4-1} = 2400\left(\dfrac{1}{2}\right)^{3} = 2400\left(\dfrac{1}{8}\right) = 300$
5	$2400\left(\dfrac{1}{2}\right)^{5-1} = 2400\left(\dfrac{1}{2}\right)^{4} = 2400\left(\dfrac{1}{16}\right) = 150$
6	$2400\left(\dfrac{1}{2}\right)^{6-1} = 2400\left(\dfrac{1}{2}\right)^{5} = 2400\left(\dfrac{1}{32}\right) = 75$

Section 2.4

15. (a) Count the number of toothpicks in each figure.

n	$T(n)$
1	4
2	12
3	20
4	28
5	36
6	44
7	52
8	60

(b) Notice the numbers have a common difference of 8 between consecutive terms. This sequence is arithmetic. Let $a = 4$ and $d = 8$.

(c) In an arithmetic sequence, $T(n) = a + (n - 1)\, d$.
Therefore, $T(n) = 4 + (n - 1)\, 8$ or $T(n) = 4 + 8n - 8 = 8n - 4$.

(d) $T(20) = 8\,(20) - 4 = 160 - 4 = 156$. $T(150) = 8\,(150) - 4 = 1200 - 4 = 1196$.

(e) The domain is the set of values for n. Therefore, the domain is $\{1, 2, 3, 4, \ldots\}$. The range is $\{4, 12, 20, 28, \ldots\}$.

Section 2.4

16. (a) Count the number of toothpicks in each figure.

n	$T(n)$
1	3
2	9
3	18
4	30
5	45
6	63
7	84
8	108

(b) There is not a common multiple nor is there a common difference between consecutive terms, so this is neither a geometric nor an arithmetic sequence.

(c) Notice that the number of triangles in the nth figure is the sum of the first n natural numbers or $\dfrac{n(n+1)}{2}$ triangles. Since there are 3 toothpicks in each triangle, we can say $T(n) = 3\dfrac{n(n+1)}{2}$.

(d) $T(15) = 3\dfrac{15(15+1)}{2} = 3\dfrac{(15)(16)}{2} = 3(15)(8) = 360.$

$T(100) = 3\dfrac{100(100+1)}{2} = 3\dfrac{(100)(101)}{2} = 3(50)(101)$
$= 15,150.$

(e) The domain is $\{1, 2, 3, 4, \dots\}$.
The range is $\{3, 9, 18, 30, 45, \dots\}$.

Section 2.4

17. (a)

n = Number of Years	Annual Interest Earned	Value of Account
0	0	100
1	5	105
2	5	110
3	5	115
4	5	120
5	5	125
6	5	130
7	5	135
8	5	140
9	5	145
10	5	150

(b) Since the difference between consecutive terms is 5 dollars, the sequence is arithmetic, with $a = 100$ and $d = 5$. (Note: In the text, terms are usually numbered beginning with 1. Here we begin numbering with 0. Therefore, the formula for the arithmetic sequence is $a + nd$.) $A(n) = 100 + n(5)$.

Section 2.4

18. (a) See the table below.

n = Number of Years	Annual Interest Earned	Value of Account
0	0	100.00
1	(0.05) (100) = 5	105.00
2	(0.05) (105) = 5.25	110.25
3	(0.05) (110.25) = 5.51	115.76
4	(0.05) (115.76) = 5.79	121.55
5	(0.05) (121.55) = 6.08	127.63
6	(0.05) (127.63) = 6.38	134.01
7	(0.05) (134.01) = 6.70	140.71
8	(0.05) (140.71) = 7.04	147.75
9	(0.05) (147.75) = 7.39	155.13
10	(0.05) (155.13) = 7.76	162.89

(b) Since the value of the account in any one year is 1.05 times the value of the account in the previous year, the sequence for the value of the account is geometric, with $a = 100$ and $r = 1.05$. So $A(n) = ar^n$, or $A(n) = 100(1.05)^n$ (Notice that we began numbering terms in the sequence with zero, so the general geometric sequence changed to ar^n)

(c) The value of the account after 10 years using simple interest is $150 (from problem 17) and the value using compound interest is $162.89. You earn $162.89 − $150.00 = $12.89 more when compounding.

Section 2.4

19. $h(1) = -16(1)^2 + 64(1) = -16 + 64 = 48$ feet.
$h(2) = -16(2)^2 + 64(2) = -16(4) + 128 = -64 + 128 = 64$ feet.
$h(3) = -16(3)^2 + 64(3) = -144 + 192 = 48$ feet.
To determine how many seconds of flight she has, we need to determine t when $h(t) = 0$ since at a height of 0 feet, she is either starting or finishing her trip.
$$0 = -16t^2 + 64t$$
$$0 = -16t (t - 4)$$
$$t = 0 \text{ or } t = 4.$$
Therefore, after 4 seconds she hits the ground. Her flight was 4 seconds long.

SOLUTIONS - PART A PROBLEMS

Chapter 3: Whole Numbers - Operations and Properties

Section 3.1

16. At this point, we only know that 1 is in the set. For the set to be closed under addition, we must be able to select any two elements, not necessarily different, add them and end up with an element in the set. Since 1 is in the set, 1 + 1 = 2 must be in the set. It follows that 1 + 2 = 3 is in the set and 1 + 3 = 4 is in the set. Continue in this way, and you will generate the set of counting numbers.

Section 3.1

17. We can use only three plus or minus signs. This means we can have only four terms with one of them being at least a three-digit number. If we keep the digits in order and consider possible three-digit numbers, then we notice that any three besides 123 will yield a number too large to get back to 100. Therefore, the other six digits must form the following two-digit numbers: 45, 67, and 89. Trial and error with the signs leads to the solution 123 − 45 − 67 + 89 = 100.

Section 3.1

18. Since the magic square will contain all the numbers from 10 to 25 exactly once, and each row has the same sum, if we add the numbers from 10 to 25 and divide by 4 (the number of rows), then we will find the sum for each row.

$$(10 + 11 + 12 + \ldots + 23 + 24 + 25) \div 4 = \frac{280}{4} = 70.$$

Consider column 2. The sum of the missing numbers must be 70 − 16 − 23 = 31. The only unused pair of numbers which add to 31 are 11 and 20. Now consider row 2. The sum of the missing numbers must be 70 − 19 − 17 = 34. The only possible pairs of unused numbers which add to 34 are 12 and 22, 13 and 21, 14 and 20. Since column 2 and row 2 contain a common number, we see that we must choose the pair 14 and 20 since 20 is also in column 2.

Fill in these numbers, placing 20 at the intersection of row 2 and column 2.

Consider column 1. The missing number must be 70 − 25 − 14 − 18 = 13.

Consider row 4. The missing number must be 70 − 13 − 23 − 10 = 24.

Consider column 3. The missing numbers must add to 70 − 19 − 24 = 27. The only unused pair of numbers that add to 27 are 12 and 15. In order to figure out which squares to place them in, we need information from another row.

25	11	12	22
14	20	19	17
18	16	15	21
13	23	24	10

Consider row 1. The missing numbers must add to 70 − 25 − 11 = 34. The only unused pair of numbers that add to 34 are 12 and 22. Since 12 must also be in column 3, place 12 at the intersection of row 1 and column 3. The final unused number, 21, must be placed at the intersection of row 3 and column 4. The final solution is shown at left.

Section 3.1
19. Notice that when the three diagonals are drawn in, a hexagon is a collection of six equilateral triangles. What is "magic" about the magic hexagon is that the sum of the numbers along each side of each equilateral triangle is the same.

Section 3.1
20. (a) Use the method given:

	(i)		(ii)		(iii)	
		39		87		32
		+93		+78		+23
		132		165		55
		231		561		
		363		726		
				627		
				1353		
				3531		
				4884		

(b) In part (a), we notice that when the sum of the digits of the original number is less than 10, the procedure takes one step because we do not carry at all. If the sum of the two digits is 15 or greater, then the procedure takes more than three steps because there would be a carry in the first sum, causing the total to exceed 150. This, in turn, causes carries in at least the next two steps. Any two-digit numbers with a sum of at least 15, such as 69, 78, 79, or 89, will take more than three steps.

Section 3.1
21. Let n = the youngest daughter's age. Since the number of years between the youngest daughter's age and the next youngest daughter's age equals the youngest daughter's age, the next youngest daughter must be $n + n = 2n$ years old. Since the daughters' ages are spaced evenly, the other daughters must be ages $3n$, $4n$ and $5n$, respectively. The oldest daughter is 16 years older than the youngest daughter, so $n + 16 = 5n$. Solving, we see that $4n = 16$ and $n = 4$. The ages are 4, 8, 12, 16, and 20.

Section 3.1

22. Consider set A. If we were to count all the elements in that set, we would include all the elements in the shaded region.

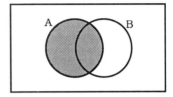

Similarly, for set B, we would count all the elements in the shaded region.

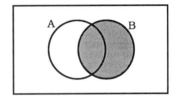

Notice that the elements in A∩B have been counted twice. We must remove this repetition when determining the number of elements in the union. Therefore, the number of elements in the union of A and B is the number of elements in A plus the number of elements in B minus the number of elements in their intersection.

Section 3.1

23. This property is true for all whole numbers. Recall that the whole numbers are {0, 1, 2, ... }. To convince your students, you might consider using a balance scale. Place objects (identical coins or other identical objects) on the scale so that it balances. Remove one item from each side and notice that the remaining items balance. Remove two items from each side and notice that the remaining items balance. Continue until you are convinced that the property will hold for all whole numbers.

Section 3.2

17. Yes, set A is closed under multiplication. For a set to be closed under multiplication, the product of any two elements (not necessarily different) of the set must be in the set. Since for the set of whole numbers multiplication can be thought of as repeated addition and the set A is closed under addition, any product, (i.e., repeated sum) is also in the set A.

Section 3.2

18. (a) To determine the sum for each row, we need to find the sum of all the numbers in the magic square. Since the sum of the numbers from 1 to 9 is 45, the total for each row must be 45/3=15. Since the numbers on the diagonal must also add to 15, the bottom left corner must be 2. Consider the row and column containing the 8. Out of the remaining digits, our only options are 6 and 1 or 4 and 3. Similarly, in the row and column containing the 2, our only options are 9 and 4 or 6 and 7. This forces the following arrangement.

$$
\begin{array}{ccc}
4 & 3 & 8 \\
9 & 5 & 1 \\
2 & 7 & 6
\end{array}
$$

(b) Let us choose a base, for example 2, and let the digits 1 through 9 be used as exponents. Since we add exponents when we multiply numbers with exponents, we can use the arrangement of numbers in the magic square from (a) to form the multiplicative magic square.

$$
\begin{array}{ccc}
2^4 & 2^3 & 2^8 \\
2^9 & 2^5 & 2^1 \\
2^2 & 2^7 & 2^6
\end{array}
$$

Each row, column and diagonal has a product of 2^{15}.

Section 3.2

19. Notice that the pattern is the line of consecutive odd numbers beginning with 1, and each line contains one more number than the previous line. The sum of the numbers in each line is the cube of the number of terms in the line. Therefore, the next three lines in the pattern must be:

$$31 + 33 + 35 + 37 + 39 + 41 = 6^3 = 216$$
$$43 + 45 + 47 + 49 + 51 + 53 + 55 = 7^3 = 343$$
$$57 + 59 + 61 + 63 + 65 + 67 + 69 + 71 = 8^3 = 512$$

Section 3.2

20. Consider multiplication and division, the most restrictive operations, first. The two numbers that are multiplied must result in a single digit answer. Our only options are $2 \times 3 = 6$ or $2 \times 4 = 8$. The only division problems resulting in a whole number are $8/4 = 2$, $8/2 = 4$, $6/2 = 3$, or $6/3 = 2$. Through trial and error, we find that $6/3 = 2$ is the only appropriate division problem. The solution is shown next.

$$9 - 5 = 4$$
$$\times$$
$$6 \div 3 = 2$$
$$||$$
$$1 + 7 = 8$$

Section 3.2

21. Use a variable. Let n = the original number. Perform the operations.

    (1) Add 10: $n + 10$
    (2) Multiply by 2: $2(n + 10) = 2n + 20$
    (3) Add 100: $2n + 20 + 100 = 2n + 120$
    (4) Divide by 2: $(2n + 120)/2 = n + 60$
    (5) Subtract the original number: $n + 60 - n = 60$

    No matter what the original number is, the result will always be 60, the number of minutes in an hour, after performing the given set of operations.

Section 3.2

22. We need to find out who owned each frog and how each frog placed. From clue 1, we know that since Michelle's frog finished ahead of Bounce and Hoppy, her frog must be either Hippy or Pounce. From clue 2, since Hippy and Hoppy tied for second, Michelle's frog must have finished first and must be Pounce. Also, since Hippy and Hoppy tied for second, Bounce must have been third. From clue 3, since Kevin and Wendy recaptured Hoppy after he escaped from his owner, we know neither of them owned Hoppy. Since Michelle owns Pounce, the only one left to own Hoppy is Jason. From clue 4, Kevin's frog earned a red ribbon for second place. Since the other second-place frog was owned by Jason, we know Kevin owns Hippy. Finally, the only frog and owner left are Wendy and Bounce. Thus, the solution is as follows:

Place	Frog	Owner
First	Pounce	Michelle
Second	Hippy	Kevin
Second	Hoppy	Jason
Third	Bounce	Wendy

Section 3.2

23. Notice there are three unknowns. Let
    p = number of people in the group,
    t = number of cups of tea each drank, and
    c = number of cakes each ate.
    The total number of cups of tea is $p \times t$, so $30 \times p \times t$ is the cost of the tea in cents. Similarly, $p \times c$ is the total number of cakes, so $50 \times p \times c$ is the cost of the cakes in

cents. Therefore, since the total bill came to 1330 cents, we know

$$30pt + 50pc = 1330$$
$$10p(3t + 5c) = 1330$$
$$p(3t + 5c) = 133.$$

We have two factors, p and $(3t + 5c)$, which when multiplied yield 133. The factors of 133 are 1 and 133 or 7 and 19. We can eliminate 1×133 since a "group" implies more than one person, and $3t + 5c \neq 1$ when t and c are whole numbers. Therefore, $p = 7$ and $3t + 5c = 19$. Guessing and checking, we find $c = 2$ and $t = 3$. Each person had three cups of tea.

Section 3.2

24. Since the five numbers are consecutive, we can think of them as approximately the same number. If we think of them as the same number, then let x be that number. Since the product of the five numbers is 15120, we have $x^5 = 15120$, so $x \approx 6.85$. Now guess five consecutive numbers close to 7 and see if their product is 15120. For example, if we try the numbers 5, 6, 7, 8, and 9, we have $5 \times 6 \times 7 \times 8 \times 9 = 15120$. Therefore, these are the desired numbers.

Section 3.2

25. When a single creature reproduces itself, it divides itself into three new creatures (not three *additional* creatures). Construct a table and compare the number of days elapsed to the number of creatures on earth.

Days Elapsed	Number of Creatures
1	$1 = 3^0$
2	$3 = 3^1$
3	$9 = 3^2$
4	$27 = 3^3$

Notice that the number of creatures on any given day is 3 raised to a power that is 1 less than the number of days elapsed. Thus, on day 30, there must be 3^{29} creatures.

Section 3.2

26. Consider the simpler problem of finding one counterfeit coin out of three coins in a single weighing. Place one coin in each pan. If they balance, then the extra coin is counterfeit. If not, the lighter coin is counterfeit. In the case of 8 coins, place 3 coins on each pan. If they balance, weigh the remaining two coins to find the counterfeit one. If three of the coins are lighter, apply the simpler problem scenario to find the counterfeit coin.

Section 3.2

27. We want to know if the property "If $ab = bc$, then $a = b$ for all whole numbers" is true. Recall the set of whole numbers is $\{0, 1, 2, 3, 4, \ldots\}$. Try a few examples. Suppose that $ac = bc$. If we let $a = 6$, $b = 11$, and $c = 0$, then $6 \times 0 = 11 \times 0$, so $0 = 0$ is true, but $6 \neq 11$. Therefore, if $c = 0$, then $ac = bc$ even if $a \neq b$. If $c \neq 0$ and $ac = bc$, then $a = b$.

Section 3.3

11. We can distribute exponentiation over a product:

$$(ab)^m = a^m \times b^m$$

Any exponent can be written as a sum of exponents, and a number raised to a sum can be broken up into a product:

$$a^{m+n} = a^m \times a^n$$

(a) $6^{10} = (2 \times 3)^{10} = 2^{10} \times 3^{10}$ and $3^{20} = 3^{10+10} = 3^{10} \times 3^{10}$. By direct comparison, $6^{10} = 2^{10} \times 3^{10} < 3^{10} \times 3^{10} = 3^{20}$.

(b) $9^9 = (3 \times 3)^9 = 3^9 \times 3^9 = 3^{18}$. By direct comparison, since $18 < 20$, $9^9 = 3^{18} < 3^{20}$.

(c) $12^{10} = (3 \times 4)^{10} = 3^{10} \times 4^{10}$ and $3^{20} = 3^{10+10} = 3^{10} \times 3^{10}$. By direct comparison, $12^{10} = 3^{10} \times 4^{10} > 3^{10} \times 3^{10} = 3^{20}$.

Section 3.3

12. Make a table to show how the price changed over the period of a few years. Here $n =$ the number of 5-year periods since 1980.

Year	Price (in dollars)	n
1980	0.25	0
1985	0.50	1
1990	1.00	2
1995	2.00	3
2000	4.00	4
2005	8.00	5
2010	16.00	6
2015	32.00	7
2020	64.00	8

(a) In 1995, the price of the candy bar would be $2.00.

(b) In 2020, the price of the candy bar would be $64.00.

(c) Notice that since the price doubles or is 2 times as large at each increase, the sequence of prices is a geometric sequence, where $a = 0.25$, and $r = 2$. Therefore, $P(n) = 0.25(2)^n$.

Section 3.3

13. Consider an example. Let $n(A) = 3$, $n(B) = 5$, and $n(C) = 6$. The fact that $3 < 5$ means that we can find a 1-1 correspondence between set A and a proper subset of set B.

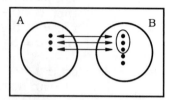

Also, $5 < 6$ means that we can find a 1-1 correspondence between set B and a proper subset of set C. Notice that the elements in A can be matched with a proper subset of the elements of B. Furthermore, this proper subset can be matched with a proper subset of the previous proper subset of C.

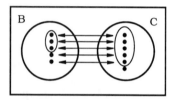

Thus, since there is a 1-1 correspondence between set A and a proper subset of C, we have $3 < 6$.

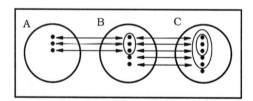

This can be generalized to the case where $a = n(A)$, $b = n(B)$, and $c = n(C)$.

Section 3.3

14. $2 + 3 + 4 = 9 = 1 + 8 = 1^3 + 2^3$.
$5 + 6 + 7 + 8 + 9 = 35 = 8 + 27 = 2^3 + 3^3$.
$10 + 11 + 12 + 13 + 14 + 15 + 16 = 91 = 27 + 64 = 3^3 + 4^3$.
(a) The next two lines in the sequence are:

$17 + 18 + 19 + 20 + 21 + 22 + 23 + 24 + 25 = 4^3 + 5^3$, and
$26 + 27 + 28 + 29 + 30 + 31 + 32 + 33 + 34 + 35 + 36 = 5^3 + 6^3$.

Notice that the number of terms in the sum is the same as the sum of the base numbers to the right of the equal sign. The first number in each sum is one more than the square of the smaller base number. The last number in the sum is the square of the larger base number.

(b) $9^3 + 10^3$ will have $9 + 10 = 19$ terms in the sum. The first term in the sum will be $9^2 + 1 = 82$, and the last term will be $10^2 = 100$. Therefore, $82 + 83 + 84 + ... + 99 + 100 = 9^3 + 10^3$.

(c) $12^3 + 13^3$ will have $12 + 13 = 25$ terms in the sum. The first term will be $12^2 + 1 = 145$, and the last term will be $13^2 = 169$. Therefore, $145 + 146 + 147 + ... + 168 + 169 = 12^3 + 13^3$.

(d) $n^3 + (n + 1)^3$ will have $n + n + 1 = 2n + 1$ terms in the sum. The first term in the sum will be $n^2 + 1$, and the last term will be $(n + 1)^2$. Therefore, we know that $(n^2 + 1) + (n^2 + 2) + (n^2 + 3) + ... + (n + 1)^2 = n^3 + (n + 1)^2$.

Section 3.3

15. Let's say the four toppings are pepperoni, onion, sausage, and anchovies. For each size pizza we decide how many toppings to put on: 0, 1, 2, 3, or 4. For each of the pizza sizes (S, M, L, XL), we may or may not include pepperoni, for example.

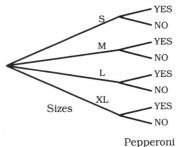

Pepperoni

At this stage, there are $4 \times 2 = 8$ possible pizza choices. For each additional topping, we may include it or not. This doubles the number of pizza types for every topping considered. Thus, there are $4 \times 2 \times 2 \times 2 \times 2 = 4 \times 2^4 = 64$ different pizza combinations.

Section 3.3

16. (a) List all possible two-digit perfect squares. Find any whose digits are also perfect squares. Keep in mind that the only one-digit perfect squares are 0, 1, 4, and 9.

<u>Two-Digit Perfect Squares</u>
16
25
36
49   Each digit is a perfect square.
64
81

49 is the only two-digit number such that it is a perfect square ($7^2 = 49$) and each of its digits is also a perfect square ($2^2 = 4$ and $3^2 = 9$).

(b) List all possible three-digit perfect squares. Find any that are made up of one one-digit perfect square and one two-digit perfect square. The lists we made from part (a) will be helpful.

Three-Digit Perfect Squares		
100	289	576
121	324	625
144	**361**	676
169	400	729
196	441	784
225	484	841
256	529	900
		961

Since 169 is made up of 16 (a two-digit square) and 9 (a one-digit square), and 361 is made up of 36 (a two-digit square) and 1 (a one-digit square), both 169 and 361 have the desired property.

For parts (c) - (f), we need a list of all four-digit perfect squares.

Four-Digit Perfect Squares				
1024	2116	3600	5476	7744
1089	2209	3721	5625	7921
1156	2304	3844	5776	8100
1225	2401	3969	5929	8281
1296	2500	4096	6084	8464
1369	2601	4225	6241	8649
1444	2704	4356	6400	8836
1521	2809	4489	6561	9025
1600	2916	4624	6724	9216
1681	3025	4761	6889	9409
1764	3136	4900	7056	9604
1849	3249	5041	7225	9801
1936	3364	5184	7396	
2025	3481	5329	7569	

(c) Use the lists we made in part (a) to find the desired numbers.

Four-Digit Square	Two One-Digit Squares	Two-Digit Square
1600	0, 0	16
1936	1, 9	36
2500	0, 0	25
3600	0, 0	36
4900	0, 0	49
6400	0, 0	64
8100	0, 0	81
9025	9, 0	25

(d) Use the list we made in part (b) to find the desired numbers.

Four-Digit Square	Three-Digit Square	One-Digit Square
1225	225	1
1444	144	4
4225	225	4
4900	900	4

(e) Use the list we made in part (a) to find the desired numbers.

Four-Digit Square	Two Two-Digit Squares
1681	16, 81

(f) Since the only single-digit perfect squares are 0, 1, 4, and 9, the only four-digit perfect squares made up of four one-digit squares are 1444, 4900, and 9409.

Section 3.3

17. Consider all possible products and compare them to perfect squares.

Mathematician One:

$2 \times 4 = 8$	$8 + 1 = 9$	$9 = 3^2$
$2 \times 12 = 24$	$24 + 1 = 25$	$25 = 5^2$
$2 \times 22 = 44$	$44 + 1 = 45$	$45 = $ Not square.

Mathematician One is incorrect.

Mathematician Two:

$2 \times 12 = 24$	$24 + 1 = 25$	$25 = 5^2$
$2 \times 24 = 48$	$48 + 1 = 49$	$49 = 7^2$
$2 \times 2380 = 4760$	$4760 + 1 = 4761$	$4761 = 69^2$
$12 \times 24 = 288$	$288 + 1 = 289$	$289 = 17^2$
$12 \times 2380 = 28560$	$28560 + 1 = 28561$	$28561 = 169^2$
$24 \times 2380 = 57120$	$57120 + 1 = 57121$	$57121 = 239^2$

Mathematician Two is correct.

Section 3.3

18. Check to see if the statements are correct.

$$10^2 - 2^2 = 100 - 4 = 96 = 12 \times 8.$$
$$20^2 - 7^2 = 400 - 49 = 351 = 27 \times 13.$$
$$80^2 - 4^2 = 6400 - 16 = 6384 = 84 \times 76.$$

To see if this works in general, we use variables and expanded form. Let ab be any two-digit number, then $ab = 10a + b$. We need to show that $(10a)^2 - b^2$ has ab, which is $10a + b$, as a factor. We know that $(10a)^2 - b^2 =$

$(10a + b)(10a - b)$.   Thus, $10a + b$ is a factor and the statement is true in general.

Section 3.3

19.   (a) Prove: If $a < b$, then $a + c < b + c$.

Suppose that $a < b$. Then we know there exists a whole number n such that $n \neq 0$ and $a + n = b$. If we add c to both sides of this equation, we have

$(a + n) + c = b + c$

$a + (n + c) = b + c$ Associative Property for Addition

$a + (c + n) = b + c$ Commutative Property for Addition

$(a + c) + n = b + c$ Associative Property for Addition

So, by the definition of less than, we see $a + c < b + c$.

(b) Prove: If $a < b$, then $a - c < b - c$.

Suppose $a < b$. Then we know there exists a whole number, n, such that $n \neq 0$, and $a + n = b$. If we subtract c from both sides of this equation, we have

$$(a + n) - c = b - c$$

$$a + (n - c) = b - c$$

$$a + (-c + n) = b - c$$

$$[a + (-c)] + n = b - c$$

$$a - c < b - c$$

SOLUTIONS - PART A PROBLEMS

Chapter 4: Whole Number Computation - Mental, Electronic, and Written

Section 4.1

28. Use rounding techniques and the guess and test method to estimate possible products.

    (a) $13 \times 6{,}000 = 78{,}000$ is too small.

    $140 \times 800 = 112{,}000$ is too small.

    $1{,}400 \times 90 = 126{,}000$ is too big, but each number was rounded up. So $1357 \times 90 = 122{,}130$.

    (b) $6 \times 70{,}000 = 420{,}000$ is too small.

    $70 \times 7{,}000 = 490{,}000$ is too big, but each number was rounded up. So check $66 \times 6{,}666 = 439{,}956$.

    (c) Notice the ones digit is a 4. Using one multiplication sign, this could only result from the 9, 6 or 4, 6 combination. So try $789 \times 3{,}456 = 2{,}726{,}784$, which is too small and $78{,}934 \times 56 = 4{,}420{,}304$, which is too big. Since we did not find the solution, we should try multiplying three numbers using two multiplication signs. Notice $8 \times 3 \times 6$ ends in a 4, so we use this as the ones digits. Therefore, $78 \times 93 \times 456 = 3{,}307{,}824$.

    (d) We use the same procedure as in (c). We see that to obtain a 5 in the ones place, we are forced to split the digits after the 5 when using only one multiplication sign. $12{,}345 \times 67 = 827{,}115$ is too big. Use two multiplication signs and notice we must break the digits at odd numbers. Therefore, $123 \times 45 \times 67 = 370{,}845$.

Section 4.1

29. To find a range for subtraction problems for three-digit numbers, we look for a way to obtain a high estimate and a low estimate. Consider two cases: borrowing or not borrowing.

    (i) When it is necessary to borrow in the hundreds place, we should round both numbers down for the high estimate. For the low estimate, round the minuend down and the subtrahend up.

    (ii) When no borrowing is needed, round the minuend up and the subtrahend down for the high estimate. Round both of them up for the low estimate.

Section 4.1

30. On most hand-held calculators, squaring a number greater than 99,999 produces an answer in scientific notation. Information is lost since not all the digits are displayed. When squaring a number greater than 99,999 it is possible to rewrite the number as a sum that can be

squared by using the distributive property. Consider the number 13,333,333. We can rewrite it as

$$13,000,000 + 333,000 + 333.$$

Therefore, $13,333,333^2$ can be written as

$(13,000,000 + 333,000 + 333)(13,000,000 + 333,000 + 333) = 169,000,000,000,000 + 4,329,000,000,000 + 4,329,000,000 + 4,329,000,000,000 + 110,889,000,000 + 110,889,000 + 4,329,000,000 + 110,889,000 + 110,889 = 177,777,768,888,889.$

Although $13,000,000 \times 13,000,000$ would be converted to scientific notation on the calculator, we can use the fact that $13 \times 13 = 169$ *will* fit in the display, and then we can add on the appropriate number of zeros.

Section 4.1

31. $99 \times 36 = 3,564$

$99 \times 23 = 2,277$

If we think of 99 as $100 - 1$, then

$99 \times 36 = (100 - 1) \times 36 = 3,600 - 36 = 3,564$, and
$99 \times 23 = (100 - 1) \times 23 = 2,300 - 23 = 2,277.$

So, $99 \times 57 = 5,700 - 57 = 5,643$, and $99 \times 63 = 6,300 - 63 = 6,237.$

Section 4.1

32. (a) Consider $25^2 = 625$, $35^2 = 1,225$, $45^2 = 2,025$, $55^2 = 3,025$. Notice that each number is of the form $a5$ where a is the tens digit and 5 is the ones digit. Squaring any number of the form $a5$ results in a number whose first digit(s) are $a(a + 1)$ and whose final digits are always 25. In general, $(a5)^2 = 100a(a + 1) + 25$. Using this method, we can find the values of

65^2: $6 \times 7 = 42$,   so $65^2 = 100(42) + 25 = 4,225.$
75^2: $7 \times 8 = 56$,   so $75^2 = 100(56) + 25 = 5,625.$
95^2: $9 \times 10 = 90$,   so $95^2 = 100(90) + 25 = 9,025.$

(b) Let $a5$ be any two-digit number, where a is the tens digit and 5 is the ones digit. Use expanded form to find the square.

$$(a5)^2 = (10a + 5)(10a + 5)$$
$$= 100a^2 + 50a + 50a + 25$$
$$= 100a^2 + 100a + 25$$
$$= 100a(a + 1) + 25$$

Section 4.1

33. George was correct. To calculate the time for the inhabitant of the moon to hear the battle of Waterloo, we need to first calculate the time in minutes required for sound to reach the moon. For sound to travel 123,256 miles at 4 miles per minute it would take $123,256/4 = 30,814$ minutes. Now convert this time to days, hours, and minutes. Since 1 day = 24 hours = 1440 minutes, 30,814 minutes is $30,814/1440 = 21$ days, with a

remainder of 574 minutes. Since 1 hour = 60 minutes, the 574 minutes is 574/60 = 9 hours, with a remainder of 34 minutes. Thus, the total time required is 21 days, 9 hours, and 34 minutes.

Section 4.1

34. Recall that when multiplying numbers, the ones digit in the solution is the same as the ones digit of the product of the ones digits of the original factors. Consider the given numbers. The ones digits are 8 and 9. Since $8 \times 9 = 72$, the ones digit of the solution is 2. This means that the given numbers could not possibly be the factors of the number consisting of 71 consecutive 1's because it has a 1 in the ones place. Since $9 \times 9 = 81$ and the ones digit is a 1, perhaps the first factor should have a 9 rather than an 8 in the ones place.

Section 4.1

35. For each product, we need to find the number that is halfway between the factors. Then we can rewrite each factor by adding to or subtracting from this number.
    (a) $54 \times 46 = (50 + 4)(50 - 4) = 50^2 - 4^2 = 2{,}484$.
    (b) $81 \times 79 = (80 + 1)(80 - 1) = 80^2 - 1^2 = 6{,}399$.
    (c) $122 \times 118 = (120 + 2)(120 - 2) = 120^2 - 2^2 = 14{,}396$.
    (d) $1{,}210 \times 1{,}190 = (1{,}200 + 10)(1{,}200 - 10) = 1{,}200^2 - 10^2 = 1{,}439{,}900$.

Section 4.1

36. Notice that the result of $898{,}423 \times 112{,}303$ will not fit into the display of the calculator. Rewrite each factor as a sum and use the distributive property: $898{,}423 \times 112{,}303 = (898{,}000 + 423)(112{,}000 + 303) = 100{,}576{,}000{,}000 + 272{,}094{,}000 + 47{,}376{,}000 + 128{,}169 = 100{,}895{,}598{,}169$.

Section 4.1

37. We can rewrite $439{,}268 \times 6{,}852$ as $(439{,}000 + 268) \times 6{,}852$ and distribute. Then we can multiply $439 \times 6{,}852$ on the eight-digit calculator and add on 3 zeros (or multiply by 1000).
    $(439{,}000 + 268) \times 6{,}852 = (439 \times 6{,}852) \times 1{,}000 + 268 \times 6{,}852 = 3{,}008{,}028 \times 1{,}000 + 1{,}836{,}336 = 3{,}008{,}028{,}000 + 1{,}836{,}336 = 3{,}009{,}864{,}336$.

Section 4.1

38. (a) If we apply the order of operations to the given expression, then we multiply first and add last. $76 \times 54 + 97 = 4{,}104 + 97 = 4{,}201$. In order to obtain 11,476, we must insert parentheses so that we add first: $76 \times (54 + 97) = 76 \times 151 = 11{,}476$.

(b) The order of operations requires that we square 13 first and then multiply by 4: $4 \times 13^2 = 4 \times 169 = 676$. In order to obtain 2,704, we need to first multiply 4×13 and then square the result: $(4 \times 13)^2 = 52^2 = 2704$.

(c) The standard order of operations produces $13 + 59^2 \times 47 = 13 + 3481 \times 47 = 13 + 163{,}607 = 163{,}620$. No parentheses are necessary.

(d) The order of operations requires that we square first and then divide, followed by subtraction and addition, respectively. $79 - 43/2 + 17^2 = 79 - 43/2 + 289 = 79 - 21.5 + 289 = 57.5 + 289 = 346.5$ Since the desired answer, 307, is not a decimal, insert parentheses to ensure that a whole number results from the division. $(79 - 43)/2 + 17^2 = (79 - 43)/2 + 289 = 36/2 + 289 = 18 + 289 = 307$.

Section 4.1

39. (a) Notice that each product is of the form $ab \times ac$, where a is the tens digit and b and c are the ones digits. The resulting product is of the form
$$a(a + 1) \times 100 + b \times c.$$
Using this form, we see that
$$57 \times 53 = 5 \times 6 \times 100 + 7 \times 3 = 3000 + 21 = 3021.$$

(b) Notice that the factors ab and ac are related further in that $b + c = 10$ so $c = 10 - b$. Thus, the factors can be written as $ab \times a(10 - b)$, where a is the tens digit, and b and $10 - b$ are the ones digits. (Note: In the expression $a(10 - b)$, the parentheses do NOT indicate multiplication but instead are used to separate the tens digit from the ones digit.)
$$\begin{aligned} ab \times a(10 - b) &= (10a + b)(10a + 10 - b) \\ &= 100a^2 + 100a - 10ab + 10ab + 10b - b^2 \\ &= 100a^2 + 100a + 10b - b^2 \\ &= 100a(a + 1) + b(10 - b) \qquad \text{Factoring} \\ &= 100a(a + 1) + bc, \text{ since } c = 10 - b. \end{aligned}$$
Thus $ab \times a(10 - b)$ has the desired form.

(c) Problem 32 is a special case where b and c are 5.

Section 4.1

40. To use the "round a 5 up" method, we must first identify the digit in the place to which we are rounding. Consider the digit to its right. If that digit is 5, 6, 7, 8, or 9, then add 1 to the digit to which we are rounding. (Note: If the digit in the place to which we are rounding is a 9 and we add 1, then we must also perform any necessary carrying.) If the digit to the right is 4 or less, then leave the digit in the place to which we are rounding alone. Finally, put zeros in all the places to the right of the digit to which we are rounding.

Section 4.2
22. The correct solution to the problem is

$$
\begin{array}{r}
1\\
29\\
+83\\
\hline
112
\end{array}
$$

Larry is not carrying properly. As a result, when he adds the 8 and the 2 (the 80 and 20), he places the 1 in the thousands place rather than in the hundreds place. Curly adds the 9 and 3 which is 12, but he carries the ones digit instead of the tens digit. Moe fails to carry the tens digit after adding 9 and 3.

Section 4.2
23. Guess and test, but look for restrictions on the sums first. In the hundreds place the numbers must total 9 or less since there is no digit in the thousands place of the solution (that is, no carry). Also, you may check for yourself to see that there must be at least one carry since each digit is used once. Two possible solutions are:

$$
\begin{array}{r}
359\\
+127\\
\hline
486
\end{array}
\qquad \text{and} \qquad
\begin{array}{r}
281\\
+673\\
\hline
954
\end{array}
$$

Section 4.2
24. (a) To obtain the greatest sum, place the largest digits in the hundreds places. The largest of the remaining digits must be placed in the tens places. The last digits are placed in the ones places.

$$
\begin{array}{r}
863\\
+742\\
\hline
1605
\end{array}
$$

(b) To obtain the least sum, place the smallest digits in the hundreds places. The smallest remaining digits must be placed in the tens places. The remaining digits are placed in the ones places.

$$
\begin{array}{r}
347\\
+268\\
\hline
615
\end{array}
$$

Section 4.2
25. There is no restriction on the number of additions that can be used. In order to add to 100, at most two digits can be paired up at a time. Since the digits must be kept in order, begin on one side and rule out possible number combinations.

(a) Begin by pairing digits from smallest to largest: 12 + 34 + 56 + 7 = 109. Since this sum is greater than 100, try to split up one of the pairs. Notice that 34 + 56 = 90, which means we still need to add 10. By splitting the number 12 into 1 + 2 and adding 7, we obtain the 10 we need. One solution is 1 + 2 + 34 + 56 + 7 = 100.

(b) If we worked with pairs from largest to smallest, we would consider 1 + 23 + 45 + 67 = 136. Notice that 23 + 67 = 90. By splitting the number 45 into 4 + 5, we obtain another solution: 1 + 23 + 4 + 5 + 67 = 100.

Section 4.2

26. In order to replace 7 digits with zeros and obtain a total of 1111, each column must add to 1, 11, or 21. Notice that the digits in the hundreds column must always add to 11.

(a) First consider the hundreds column since it must add to 11. This total might result after carrying 2, 1, or 0 from the tens column. Make a list of the possible sums of the digits in the hundreds column involving each of the carries.

If we carry 2 from the tens column, we need to find combinations of digits that add to 9. The choices are 9 + 0 + 0 + 0 + 0 or 0 + 0 + 5 + 3 + 1.
If we carry 1 from the tens column, we need to find combinations of digits that add to 10. The choices are 9 + 0 + 0 + 0 + 1 or 0 + 7 + 0 + 3 + 0.
If we carry 0 from the tens column, then the combination of digits must add to 11. The only choice is 0 + 7 + 0 + 3 + 1.

Try one of these possibilities and continue. For example, use 9 + 0 + 0 + 0 + 0. The tens column must add to 21. Make a list of the possible tens digit combinations that total 21 after possible carries of 2, 1, or 0 from the ones column.

If we carry 2 from the ones column, find combinations of digits to add to 19. Use 9 + 7 + 0 + 3 + 0. Notice that we have replaced 6 digits with zeros. We can only replace one more digit with a zero. It is impossible to replace one digit in the ones column and total 21.
If we carry 1 from the ones column, find combinations of digits that add to 20. Use 9 + 7 + 0 + 3 + 1. 5 digits have been replaced with zeros. We need to replace 2 digits in the ones column with zeros and also obtain a total of 11. Use 0 + 7 + 0 + 3 + 1.

Therefore, the solution is:

```
  990
  077
  000
  033
 +011
 1111
```

(b) Part (i). In this case, we need to replace eight digits with zeros. Notice that 999 + 111 = 1110 is close to the desired total of 1111, but nine digits have been replaced with zeros. Since the total is off by 1, consider the ones column. By replacing only two digits with zeros, we need to obtain a total of 11. We try 0 + 7 + 0 + 3 + 1. The solution is:

```
  990
  007
  000
  003
 +111
 1111
```

Part (iii). In this case, we must replace ten digits with zeros. Consider the possible ways to obtain totals of 1, 11, or 21 in the ones column.

```
 1:  0 + 0 + 0 + 0 + 1
11:  0 + 7 + 0 + 3 + 1
21:  9 + 7 + 5 + 0 + 0.
```

Since we need to replace so many digits with zeros, try the case that uses the greatest number of zeros for the ones column.

```
   0
   0
   0
   0
  +1
```

Consider the tens column. We must obtain a total of 1, 11, or 21 without any carries from the ones column. Since we still must replace six more digits with zeros, use the case with the greatest number of zeros for the tens column.

```
  00
  00
  00
  00
 +11
```

Is it possible to obtain a total of 1111 by replacing only two digits in the hundreds column with zeros? Yes, if we use 0 + 7 + 0 + 3 + 1.

```
  000
  700
  000
  300
 +111
 1111
```

Part (ii). Since we want to replace nine digits with zeros, see if it is possible to change part (iii) slightly. In the hundreds column, is it possible to obtain a total of 11 by replacing one digit with a zero? No, so consider the tens column. Try using $0 + 7 + 0 + 3 + 1$. Then in the hundreds column we must replace three digits and obtain a total of 10 since there is a carry from the tens column. Use $0 + 7 + 0 + 3 + 0$. One solution is:

```
   000
   770
   000
   330
 + 011
  1111
```

Section 4.2
27.   (a) $26 + 37 = 63$ and $36 + 27 = 63$. Notice that the sum is the same in each case.

    (b) $37 - 26 = 11$ and $36 - 27 = 9$. Notice that the answers differ by 2.

    (c) If we pick any such pairs from two consecutive columns, then the sums are always the same, and when we subtract, the answers always differ by 2. Consider several examples: For the pairs 21, 52 and 51, 22, we have $21 + 52 = 73 = 51 + 22$. We also have $52 - 21 = 31$ and $51 - 22 = 29$. For the pairs 4, 45 and 44, 5, we have $4 + 45 = 49 = 44 + 5$. Also, $45 - 4 = 41$ and $44 - 5 = 39$.

    (d) Since we see patterns for addition and subtraction, consider possible patterns for multiplication and division. Note, however, that division often results in fractions, so we consider only multiplication. For the given pairs 26, 37 and 36, 27, we have $26 \times 37 = 962$ and $36 \times 27 = 972$. Notice that there is a difference of 10 and that the tens digits in each pair differ by 1. For the pairs 21, 52 and 51, 22, we have $21 \times 52 = 1092$ and $51 \times 22 = 1122$. Notice that the products differ by 30 and that the tens digits in each pair differ by 3. Test this pattern on other pairs. For example, for the pairs 4, 45 and 44, 5, we have $4 \times 45 = 180$ and $44 \times 5 = 220$. The products differ by 40 and the tens digits in each pair differ by 4. Therefore, the difference of the products is always 10 times the difference in the tens places.

Section 4.2

28. (a) Our figure must be a 4 × 5 array of X's.

X	X	X	X	X
X	X	X	X	X
X	X	X	X	X
X	X	X	X	X

(b) The sum of the X's in half of the figure can be expressed in a similar way.

$$1 + 2 + 3 + 4 = \frac{1}{2}(4 \times 5)$$
$$10 = 10.$$

The sum is correct.
Notice that the sum of the first n natural numbers is $\frac{1}{2}n(n + 1)$.

(c) $1 + 2 + ... + 50 = \frac{1}{2}(50 \times 51) = 1275.$

$1 + 2 + ... + 75 = \frac{1}{2}(75 \times 76) = 2850.$

Section 4.2

29. Notice that only seven of the ten digits are used in each problem. Since none of the digits in the total are the same, we must carry in each column *and* each carry must be different. (If we did not carry in the ones column, we would not carry in *any* column, and a 4-digit solution would not result.) Since adding any three different digits will result in a sum between 3 and 24, the only possible carries are 1 and 2. Consider the ones column. In order to carry a 1 into the tens column *and* carry a 2 into the hundreds column, the ones column must total 19. This forces the tens column to total 20 and the hundreds column to total 21. So RSTU must equal 2109. All possible digit combinations which add to 19 without repeating digits are 8 + 7 + 4 or 8 + 6 + 5. Our two solutions are:

888	888
777	666
+444	+555
2109	2109

Section 4.2

30. Bob: Although Bob does not show his carries, he seems to add them correctly. His problem is that he only multiplies ones digits with ones digits, tens digits with tens digits, hundreds digits with hundreds, and so on. For the last question, Bob's solution would be:

$$
\begin{array}{r}
84 \\
\times\,26 \\
\hline
184
\end{array}
$$

Jennifer: Notice that she never includes the carries in the sum, although she does carry correctly. For the last question, Jennifer's solution would be:

$$
\begin{array}{r}
3 \\
25 \\
\times\,6 \\
\hline
120
\end{array}
$$

Suzie: Suzie carries correctly, however she adds the tens digit and carry before multiplying. Her solution for the last question would be:

$$
\begin{array}{r}
3 \\
29 \\
\times\,4 \\
\hline
206
\end{array}
$$

Tom: Any multiplication by a single digit number is done correctly. Tom does show his carries correctly. However, he has the same problem that Bob has. Tom's final solution would look like:

$$
\begin{array}{r}
2 \\
517 \\
\times\,463 \\
\hline
2081
\end{array}
$$

Consider algorithms and models that might help each student overcome his/her problems. Bob and Tom might benefit from using expanded form and the distributive property. This would force them to remember to multiply each digit in the first number by each digit in the second number no matter what their place value. Jennifer might benefit from using a concrete model that would force her to add in the carries. Suzie needs to use one of the intermediate algorithms so that she is forced to multiply the digits in the factors first and then add the carry.

Section 4.2

31. In this algorithm it is necessary to locate each product in the correct place value position. To do this, recall some general multiplication facts.

When multiplying digits in the following place values	Place the last digit of the number in the
ones × tens	tens place
ones × hundreds	hundreds place
ones × thousands	thousands place
tens × tens	hundreds place
tens × hundreds	thousands place
tens × thousands	ten thousands place

In the algorithm, the digits in one factor are systematically multiplied by the digits in the other factor.

		Product
Step 1:	**35** 496**7**	21 (1 in the tens place)
Step 2:	**35** 49**6**7	1835 (8 in the hundreds place) (5 in the ones place)
Step 3:	**35** 4**9**67	2730 (7 in the thousands place) (0 in the tens place)
Step 4:	**35** **4**967	1245 (2 in the ten thousands place) (5 in the hundreds place)
Step 5:	3**5** **4**967	20 (0 in the thousands place)

Therefore, we can calculate the product 4967 × 35 as shown to the left.

```
      2 1
    1 8 3 5
    2 7 3 0
  1 2 4 5
    2 0
  _____
  1 7 3 8 4 5
```

Using the same steps, we can multiply 5314 and 79.

		Product
Step 1:	**79** 531**4**	28 (8 in the tens place)
Step 2:	**79** 53**1**4	736 (7 in the hundreds place) (6 in the ones place)
Step 3:	**79** 5**3**14	2109 (1 in the thousands place) (9 in the tens place)
Step 4:	**79** **5**314	3527 (5 in the ten thousands place) (7 in the hundreds place)
Step 5:	7**9** **5**314	45 (5 in the thousands place)

Therefore, we can calculate the product 5314×79 as follows:

$$
\begin{array}{r}
2\,8 \\
7\,3\,6 \\
2\,1\,0\,9 \\
3\,5\,2\,7 \\
4\,5 \\
\hline
4\,1\,9\,8\,0\,6
\end{array}
$$

Section 4.3

9. Notice that the smallest base in which this addition problem could be done is base 6, since the digits 1 through 5 are used. Consider the sum of the digits in the units places. $5 + 2$ yields 1 in the units place of the sum. This would not happen in base 8 or a base larger than 8. In a base of at least 8, we would have $5 + 2 = 7$. If the problem was done in base 7, we would have $5 + 2 = 10$, so there would be a 0 in the units place of the sum. Therefore, this problem must have been done in base 6. Alternately, consider base blocks. In that case, we would have 5 units + 2 units = 1 long + 1 unit in base 6.

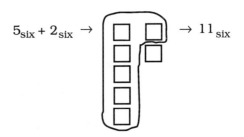

$$5_{\text{six}} + 2_{\text{six}} \rightarrow \qquad \rightarrow 11_{\text{six}}$$

Section 4.3

10. Steve has the least amount of money. We can find out how much money everyone has if we know how much Steve has. Use a variable. Let $x =$ the amount of money Steve has. Since Tricia has \$21 more than Steve, Tricia has $x + 21$ dollars. Since Bill has \$17 more than Tricia, Bill has $x + 21 + 17$ dollars. Since Jane has \$10 more than Bill, Jane has $x + 21 + 17 + 10$. All of their money together is \$115 so we have

$$115 = x + (x + 21) + (x + 21 + 17) + (x + 21 + 17 + 10)$$
$$115 = 4x + 107$$
$$8 = 4x$$
$$2 = x.$$

Therefore, Steve has \$2, Tricia has \$23, Bill has \$40, and Jane has \$50.

Section 4.3

11. To determine which of the numbers is a perfect square, consider the digit in the ones place of each number. Could that digit be the result of a number times itself? To answer that question, notice that when any number is squared, the ones digit is always the same as the ones digit of one of the following: 0^2, 1^2, 2^2, 3^2, 4^2, 5^2, 6^2, 7^2, 8^2, or 9^2 (i.e., the ones digit of the square of the ones digit). In general, when multiplying numbers if we are only interested in the resulting ones digit, we do not need to calculate the whole product. It is only necessary to look at the product of the digits in the ones places. The units digit in that product is the ones digit in the original product.

Possible Ones Digits	Square	Ones Digit of Square
0	0	0
1	1	1
2	4	4
3	9	9
4	16	6
5	25	5
6	36	6
7	49	9
8	64	4
9	81	1

For a perfect square, the only digits that could be in the ones place are 0, 1, 4, 5, 6, or 9. Therefore, 39,037,066,084 is the only perfect square in the list.

Section 4.3

12. Make a list of the perfect squares greater than 100. Calculate what number had to be added to 100 in each case to obtain the perfect square. Add that number to 164 and see if a new perfect square is created.

Perfect Square	Square − 100	(Square − 100) + 164
121	21	185 not a square
144	44	208 not a square
169	69	233 not a square
196	96	260 not a square
225	125	$289 = 17^2$

If 125 is added to 100 and to 164, the results are 225 and 289. Both are perfect squares.

SOLUTIONS - PART A PROBLEMS
Chapter 5: Number Theory

Section 5.1
14. Divisibility by 5: If abc is any three-digit number, then let $n = a \times 10^2 + b \times 10 + c$ be the number in expanded form. Observe that $a \times 10^2 + b \times 10 = 10(10a + b)$. Since $5|10$, it follows that $5|10(a \times 10 + b)$ or $5|(a \times 10^2 + b \times 10)$ for any digits a and b. So, if $5|c$, then $5|[10(a \times 10 + b) + c]$, and $5|(a \times 10^2 + b \times 10 + c)$. Thus, $5|n$. Conversely, let $5|(a \times 10^2 + b \times 10 + c)$. Since $5|(a \times 10^2 + b \times 10)$, we know $5|[(a \times 10^2 + b \times 10 + c) - (a \times 10^2 + b \times 10)]$. Thus, $5|c$.

Divisibility by 10: If abc is any three-digit number, then let $n = a \times 10^2 + b \times 10 + c$ be the number in expanded form. Observe that $a \times 10^2 + b \times 10 = 10(a \times 10 + b)$. Since $10|10$, we know $10|10(a \times 10 + b)$, or $10|(a \times 10^2 + b \times 10)$ for any digits a and b. So, if $10|c$, then we know $10|[10(a \times 10 + b) + c]$, or $10|(a \times 10^2 + b \times 10 + c)$. Thus, $10|n$. Conversely, let $10|(a \times 10^2 + b \times 10 + c)$. Since $10|(a \times 10^2 + b \times 10)$, it follows that $10|[(a \times 10^2 + b \times 10 + c) - (a \times 10^2 + b \times 10)]$. Therefore, $10|c$. (Note: For 10 to divide c, a single-digit number, c would have to be zero.)

Section 5.1
15. (a) True. Since $6! = 6 \times 5 \times 4 \times 3 \times 2 \times 1$ has 6 as a factor, $6|6!$.
    (b) True. Since $6! = 6 \times 5 \times 4 \times 3 \times 2 \times 1$ has 5 as a factor, $5|6!$.
    (c) False. $6!$ does not contain 11 as a factor and 11 is a prime. Therefore, it cannot be the result of some product of factors of $6!$.
    (d) True. Since $30! = 30 \times 29 \times ... \times 3 \times 2 \times 1$ has 30 as a factor, $30|30!$.
    (e) True. Since $40 = 10 \times 4$ and $30! = 30 \times 29 \times ... \times 11 \times 10 \times 9 \times ... \times 4 \times 3 \times 2 \times 1$, we see $30!$ contains 10 and 4 as factors. Therefore $40|30!$.
    (f) $30!$ has 5 as a factor so $30!$ must end in a 0 or 5. It follows that $30! + 1$ must end in 1 or 6. Notice that 30 has 5 as a factor. Therefore, 30 cannot be divisor of $30! + 1$ since $30! + 1$ does not end in a 0 or 5.

Section 5.1

16. (a) It is true that 8 divides 7!. Notice that $8 = 2 \times 4$, and $7! = 7 \times 6 \times 5 \times 4 \times 3 \times 2 \times 1$. 7! contains both 2 and 4 as factors. Therefore, 7! contains 8 as a factor.

    (b) False. 7 does not divide 6!. Since 7 is a prime, it cannot be the result of a product of factors of 6!. Therefore, 6! does not have 7 as a factor.

    (c) Notice how the numbers are related in parts (a) and (b). In part (a), 8 is a composite number and can be written as a product of factors of 7!. In part (b), 7 is a prime number and cannot be written as a product of factors of 6!. In general, if n is a composite number greater than 4, then n divides $(n - 1)!$. If n is prime, then n does not divide $(n - 1)!$.

Section 5.1

17. If the one composite number in the list has a factor less than 30, we know we only need to check the primes 2, 3, 5, 7, 11, 13, 17, 19, 23, and 29 to see if they are factors of the number. No number in the set is even, so no number has a factor of 2. Since each number consists of 3's and a 1, the total of the digits cannot be divisible by 3. Therefore, none has a factor of 3. No number ends in 0 or 5, so no number has a factor of 5. Use a calculator to check for divisibility by 7, 11, ..., 29. The composite number is $333,333,331 = 17(19607843)$.

Section 5.1

18. $P(n) = n^2 + n + 17$

n	$P(n)$	
0	17	Prime
1	19	Prime
2	23	Prime
3	29	Prime
4	37	Prime
5	47	Prime
6	59	Prime
7	73	Prime
8	89	Prime
9	107	Prime
10	127	Prime
11	149	Prime
12	173	Prime
13	199	Prime

Continue the list until P(n) yields a composite number. Since P(17) = 17^2 + 17 + 17 = 17 × 19, we know we need only try whole numbers up to 17.

14	227	Prime
15	257	Prime
16	289 = 17 × 17	Composite

16 is the smallest whole number for which P(n) yields a composite number.

Section 5.1

19. (a)

n	$n^2 + n + 41$
0	41
1	43
2	47
3	53
4	61
5	71
6	83
7	97
8	113
9	131
10	151

All the numbers are prime. Using divisibility tests, it is only necessary to check primes up to 13 to see if 151 is prime.

(b) Consider the following table.

131	130	129	128	127	126	125	124	123	122	
132	97	96	95	94	93	92	91	90	121	
133	98	71	70	69	68	67	66	89	120	
134	99	72	53	52	51	50	65	88	119	
135	100	73	54	43	42	49	64	87	118	
136	101	74	55	44	41	48	63	86	117	
137	102	75	56	45	46	47	62	85	116	
138	103	76	57	58	59	60	61	84	115	
139	104	77	78	79	80	81	82	83	114	
140	105	106	107	108	109	110	111	112	113	
141	142	143	144	145	146	147	148	149	150	151

All the prime numbers generated from the formula in part (a) are in the main diagonal (upper left to lower right).

Section 5.1

20. Recall the divisibility test for 11. Notice that any "repunit" number consisting of an even number of ones will be divisible by 11. Factor these numbers and look for a pattern.

	Number	Factors
2 1's	11	11×1
4 1's	1111	11×101
6 1's	111111	11×10101
8 1's	11111111	11×1010101
10 1's	1111111111	11×101010101
12 1's		11×10101010101
14 1's		11×1010101010101
16 1's		$11 \times 101010101010101$
18 1's		$11 \times 10101010101010101$

Each of these "repunit" numbers is divisible by 11. Notice that the other factor always contains half as many ones as the "repunit" number. Next, recall the divisibility test for 3. Any "repunit" number made up of 3 ones or a multiple of 3 ones will be divisible by 3. Factor these numbers and look for a pattern.

	Number	Factors
3 1's	111	3×37
6 1's	111111	3×37037
9 1's	111111111	3×37037037
12 1's		3×37037037037
15 1's		3×37037037037037

Each of these "repunit" numbers is divisible by 3. Notice that the other factor contains one less digit than the "repunit" number, repeats the combination 370, and ends in 37. The only "repunit" numbers left to factor are those with 5, 7, 11, 13, and 17 ones.

	Number	Factors
5 1's	11111	41×271
7 1's	1111111	239×4649
11 1's	11111111111	21649×513239
13 1's		$53 \times 79 \times 265371653$
17 1's		$2071723 \times 5363222357$

Section 5.1

21. The only number in the list that can be written as the sum of two primes is 7: 7 = 2 + 5. Notice all the numbers in the list are odd, and the sum of an odd number and an even number is always odd. Since the only even prime is 2, all the numbers must be the result of adding 2 and some other prime. If 2 is subtracted from each number in the list, the resulting number ends in 5. With the exception of the number 5 itself, any number ending in 5 will not be a prime.

Section 5.1

22. (a) List prime numbers less than 100 and note which primes are one more than a multiple of 4.

Prime	Prime
2	43
3	47
5 = 4(1) + 1	53 = 4(13) + 1
7	59
11	61 = 4(15) + 1
13 = 4(3) + 1	67
17 = 4(4) + 1	71
19	73 = 4(18) + 1
23	79
29 = 4(7) + 1	83
31	89 = 4(22) + 1
37 = 4(9) + 1	97 = 4(24) + 1
41 = 4(10) + 1	

(b) Since each prime of the form $4x + 1$ where x is a whole number is to be written as the sum of two square numbers, it may be helpful to list the perfect squares less than 100. Perfect squares: 1, 4, 9, 16, 25, 36, 49, 64, 81.

$$5 = 4 + 1$$
$$13 = 9 + 4$$
$$17 = 16 + 1$$
$$29 = 25 + 4$$
$$37 = 36 + 1$$
$$41 = 25 + 16$$
$$53 = 49 + 4$$
$$61 = 25 + 36$$
$$73 = 64 + 9$$
$$89 = 64 + 25$$
$$97 = 81 + 16$$

Section 5.1

23. No. In each pair of consecutive whole numbers, one is even and one is odd. Any even number is divisible by 2. Therefore, with the exception of the number 2, no even number can be prime. The whole numbers 2 and 3 are the only consecutive primes.

Section 5.1

24. Eliminate possible pairs of numbers by recalling that any even number (except 2) is composite. Also, any number ending in 5 (except 5) is composite, so it would not be paired with a possible prime number ending in 3 or 7. The only pairs of numbers to consider would be primes whose ones digits are 1 and 3, 7 and 9, 9 and 1. Make a list of these pairs of numbers. Many pairs can be eliminated quickly by using divisibility tests to show one or both numbers in the pair are composite. The twin primes less than 200 are 3 and 5, 17 and 19, 41 and 43, 59 and 61, 71 and 73, 101 and 103, 107 and 109, 137 and 139, 149 and 151, 179 and 181, 191 and 193, 197 and 199.

Section 5.1

25. (a)

$14 = 3 + 11$	$26 = 3 + 23$	$38 = 7 + 31$
$16 = 5 + 11$	$28 = 5 + 23$	$40 = 3 + 37$
$18 = 7 + 11$	$30 = 7 + 23$	
$20 = 3 + 17$	$32 = 3 + 29$	
$22 = 5 + 17$	$34 = 5 + 29$	
$24 = 7 + 17$	$36 = 7 + 29$	

(b) Let n be any odd number greater than 6. Consider the prime number 3. Because $n - 3$ must be an even number, $n - 3$ can be expressed as the sum of two prime numbers by Goldbach's conjecture. If p and q are prime numbers, then $n - 3 = p + q$. Therefore, we have that $n = 3 + p + q$, so any odd number can be expressed as the sum of three primes.

Section 5.1

26. It may be helpful to make a list of primes. When a pair of primes is found between a number and its double, see if that pair of primes falls between other numbers and their doubles. The primes 7 and 11 fall between 6 and its double 12. Consider the primes 11 and 13. Each of the numbers 7, 8, 9 and 10 falls below 11, and each double, 14, 16, 18, and 20 is larger than 13. Consider the primes 29 and 31. Each of the numbers 17 through 28 is less than 29, and each double is greater than 31. The primes

31 and 37 fall between 29 and 30 and their doubles. Finally, for every number from 31 to 49, its double is greater than 60 so the primes 53 and 59 fall in between.

Section 5.1
27. Factor 1,234,567,890 into primes. We find that $1,234,567,890 = 2 \times 3^2 \times 5 \times 3607 \times 3803$. In order to obtain two factors as close together as possible, multiply 3607 and 3803 by some combination of the remaining factors. The remaining factors are $2 \times 3^2 \times 5$. Notice that $2 \times 5 = 10$ and $3^2 = 9$. If we multiply $3607 \times 10 = 36070$ and $3803 \times 9 = 34227$, then the two factors with the smallest difference whose product is 1,234,567,890 are 36070 and 34227.

Section 5.1
28. (a) Six divides every number in the set. Notice that every number in the set is formed from the product of three consecutive counting numbers. Of the three consecutive counting numbers, at least one is even (has a factor of 2) and one is a multiple of 3. Therefore, since $2 \times 3 = 6$, we know 6 divides each number in the set. Since the smallest number in the set is $1 \times 2 \times 3 = 6$, no number larger than 6 could divide each number.

(b) Three divides every number in the set. Notice that each number in the set is formed from the product of three consecutive odd numbers or three consecutive even numbers. If n is any counting number, then each number in the set is of the form $n(n + 2)(n + 4)$. When any number is divided by 3, the only possible remainders are 0, 1, and 2. Consider n. If n divided by 3 has a remainder of 0, then it is a multiple of 3, and the product has a factor of 3. If n divided by 3 has a remainder of 1, then $n + 2$ must be divisible by 3. If the remainder is 2, then $n + 4$ is divisible by 3. Therefore, one of the factors of each number in the set is divisible by 3.

Section 5.1
29. The smallest counting number that is divisible by the numbers 2, 3, 4, 5, 6, 7, 8, 9, and 10 must contain each of them as a factor. Consider the prime factorization of each number: 2, 3, 2^2, 5, 2×3, 7, 2^3, 3^2, 2×5. The desired number must contain at least 3 factors of 2, 2 factors of 3, 1 factor of 5, and 1 factor of 7. Therefore, the smallest

counting number that is divisible by the numbers 2 through 10 is $2^3 \times 3^2 \times 5 \times 7 = 2520$.

Section 5.1

30. The smallest counting number that is divisible by 2, 4, 5, 6, and 12 must have each number as a factor. Consider the prime factorization of each number: 2, 2^2, 5, 2×3, $2^2 \times 3$. The desired number must contain at least 2 factors of 2, 1 factor of 3, and 1 factor of 5. The smallest number divisible by 2, 4, 5, 6, and 12 is $2^2 \times 3 \times 5 = 60$.

Section 5.1

31. Let n be any counting number. Then n, $n+1$, and $n+2$ are any three consecutive counting numbers. The sum of these numbers is $n + (n + 1) + (n + 2) = 3n + 3 = 3(n + 1)$. Notice that 3 is a factor of the sum. Therefore, the sum of three consecutive counting numbers always has a divisor of 3.

Section 5.1

32. Try some examples:

$$3, 8, 11, 19, 30, 49, \underline{79}, 128, 207, 335$$

The sum is $869 = 11(79)$, and 79 is the seventh number.

$$17, 28, 45, 73, 118, 191, \underline{309}, 500, 809, 1309$$

Here the sum is $3399 = 11(309)$, and 309 is the seventh number.

Let a and b be any two counting numbers. If a sequence of ten numbers is formed so that each new term is the sum of the two preceding numbers, then the seventh number is a factor of the sum. That is, if the ten numbers are a, b, $a + b$, $a + 2b$, $2a + 3b$, $3a + 5b$, $5a + 8b$, $8a + 13b$, $13a + 21b$, and $21a + 34b$, then $5a + 8b$ is the seventh term. The sum is $55a + 88b = 11(5a + 8b) = 11$(seventh number).

Section 5.1

33. (a) Recall that for whole numbers a, m, and n, where $a \neq 0$, if $a|m$ and $a|n$ then $a|(m + n)$. Since $2|5!$ and $2|2$ then $2|(5! + 2)$. Therefore, $5! + 2$ is a composite number. Similarly, $5! + 3$, $5! + 4$, and $5! + 5$ are also composite.

(b) Note that by using 5!, we obtained 4 consecutive composite numbers. If we want 1000 consecutive composite numbers, then we should use 1001! and follow the example set in (a). The numbers $1001! + 2$, $1001! + 3$, $1001! + 4$, ..., $1001! + 1001$ are all composite numbers. If m is a counting number and $2 \leq m \leq 1001$, then $m \mid 1001!$ and $m \mid m$, so $m \mid (1001! + m)$.

Section 5.1

34. Calculate the cost of the apples and potatoes. The apples cost 5×27 cents = 135 cents = \$1.35. The potatoes cost 2×78 cents = 156 cents = \$1.56. Together they cost \$1.35 + \$1.56 = \$2.91. Since the total cost calculated by the cashier was \$3.52, the cantaloupes and lemons must have cost \$3.52 − \$2.91 = \$0.61. If c = cost of one cantaloupe and l = the cost of one lemon, then $3c + 6l = 61$ cents.

$$3(c + 2l) = 61$$

Notice that 3 is a factor of the cost of the fruit, but 61 is not divisible by 3. Since nothing costs a fraction of a cent, the cashier must have made a mistake.

Section 5.1

35. Let n be the three-digit number. Then $7 \mid (n − 7)$, $8 \mid (n − 8)$, and $9 \mid (n − 9)$. Since $7 \mid 7$ and $7 \mid (n − 7)$, we know $7 \mid (n − 7 + 7)$, so $7 \mid n$. Similarly, $8 \mid n$ and $9 \mid n$. Therefore, n is divisible by 7, 8, and 9. The only three-digit number that has 7, 8, and 9 as factors is $7 \times 8 \times 9 = 504$.

Section 5.1

36. There is always one cupcake left over when cupcakes are arranged in groups of 2, 3, 4, 5, or 6. If n is the number of cupcakes, then since $n − 1$ of the cupcakes can be arranged in groups of 2 we know that $n − 1$ is a multiple of 2. We know then that $2 \mid (n − 1)$. Similarly $3 \mid (n − 1)$, $4 \mid (n − 1)$, $5 \mid (n − 1)$, and $6 \mid (n − 1)$. We want one number, $n − 1$, that contains 2, 3, 4, 5, and 6 as factors. Consider the prime factorization of each number: 2, 3, 2^2, 5, and 2×3. The smallest number containing each number as a factor must contain at least 2^2, 3, and 5. Therefore, $n − 1 = 2^2 \times 3 \times 5 = 60$ and $n = 61$ cupcakes.

Section 5.1

37. Consider any number of the form abc,abc, where a, b, and c are whole numbers. In expanded form, the number is $100,000a + 10,000b + 1000c + 100a + 10b + c = 100,100a + 10,010b + 1001c = 1001(100a + 10b + c) = 11 \times 13 \times 7(100a + 10b + c)$.

13 is a factor of this number, so any number of the form abc,abc is divisible by 13. The numbers 7 and 11 are also factors of abc,abc.

Section 5.1

38. (a) Let $abba$ be any four-digit palindrome. Write $abba$ in expanded form, simplify, and then factor.

$$\begin{aligned} abba &= 1000a + 100b + 10b + a \\ &= 1001a + 110b \\ &= 11(91a + 10 b) \end{aligned}$$

Therefore, since 11 is a factor of $abba$, any four-digit palindrome is divisible by 11.

(b) Consider the following palindromes. Notice that the only ones that are divisible by 11 are those that have an even number of digits.

Palindrome	
11	$= 11 \times 1$
101	not divisible by 11
1001	$= 11 \times 91$
10001	not divisible by 11
100001	$= 11 \times 9091$
1000001	not divisible by 11
10000001	$= 11 \times 909091$

Any palindrome with 1's at both ends and an even number of 0's between them is divisible by 11.
Consider an eight-digit palindrome written in expanded form:

$69433496 = 6(10000001) + 9(1000010) + 4(100100) + 3(11000)$

Notice that each term in the sum is a multiple of a palindrome with 1's at the ends and an even number of 0's in the middle.

$69433496 = 6(10000001) + 9 \times 10(100001) + 4 \times 100(1001) + 3 \times 1000(11)$

Since 10000001, 100001, 1001, and 11 are all divisible by 11, the sum is divisible by 11. In general, any palindrome of the form $abc \cdots cba$ with an even number of digits will be divisible by 11. You can prove this, as we did in our example, by writing the palindrome in expanded form as a combination of terms involving known multiples of 11.

Section 5.1

39. Let c = the cost of one calculator then $c \mid 2567$ and $c \mid 4267$. Consider the theorem which states that for whole numbers a, m, n, where $a \neq 0$ and $m \geq n$, if $a \mid m$ and $a \mid n$, then $a \mid (m - n)$. By this theorem we know that $c \mid (4267 - 2567)$, so $c \mid 1700$. (Notice that $1700 = 2^2 \times 5^2 \times 17$.) Since neither 2567 nor 4267 is even or ends in 0 or 5, neither number contains 2 or 5 as a factor. The only other number that could possibly divide 2567, 4267, and 1700 is 17. Therefore, the cost of one calculator is $17. In the first year, $2567/17 = 151$ calculators were sold. In the second year, $4267/17 = 251$ calculators were sold.

Section 5.1

40. Notice that $7 \mid 2149$ since $2149 = 7 \times 307$, and $7 \mid 149{,}002$ since $149{,}002 = 7 \times 21{,}286$. Try some examples.

$$7 \mid 7231 \text{ and } 7 \mid 231{,}007$$
$$7 \mid 5964 \text{ and } 7 \mid 964{,}005$$

Conjecture: If $7 \mid abcd$, then $7 \mid bcd00a$.

Proof: Suppose $7 \mid abcd$. Add $abcd$ and $bcd00a$ in expanded form and simplify: $(1000a + 100b + 10c + d) + (100{,}000b + 10{,}000c + 1000d + a) = 1001a + 100{,}100b + 10{,}010c + 1001d = 7 \times 143(a + 100b + 10c + d)$.
Therefore, $7 \mid (abcd + bcd00a)$. Since $7 \mid abcd$ and $7 \mid (abcd + bcd00a)$, then $7 \mid [(abcd + bcd00a) - (abcd)]$, so $7 \mid bcd00a$.

Section 5.1

41. Notice that the first number in each sum is the total from the previous equation. Continue the list by adding the next consecutive even number to the total from the previous equation, as shown next.

$$37 + 10 = 47$$
$$47 + 12 = 59$$
$$59 + 14 = 73$$
$$73 + 16 = 89$$
$$89 + 18 = 107$$
$$107 + 20 = 127$$
$$127 + 22 = 149$$
$$149 + 24 = 173$$
$$173 + 26 = 199$$
$$199 + 28 = 227$$
$$227 + 30 = 257$$
$$257 + 32 = 289 = 17^2 \quad \text{not a prime}$$

Section 5.1

42. Let $a \times 10^3 + b \times 10^2 + c \times 10 + d$ be any four-digit number. Then

$$
\begin{aligned}
a \times 10^3 + b \times 10^2 + c \times 10 + d &= a(1001 - 1) + b(99 + 1) + c(11 - 1) + d \\
&= 1001a - a + 99b + b + 11c - c + d \\
&= 1001a + 99b + 11c - a + b - c + d \\
&= 11(91a + 9b + c) - a + b - c + d \\
&= 11(91a + 9b + c) - (a - b + c - d).
\end{aligned}
$$

Notice that $11 \mid 11(91a + 9b + c)$. For 11 to divide $abcd$, 11 must also divide $a - b + c - d$. This corresponds to the test for divisibility for 11. A number is divisible by 11 if and only if 11 divides the difference of the sum of the digits whose place values are odd powers of 10 and the sum of the digits whose place values are even powers of 10. In $abcd$, a and c are in the place values 10^3 and 10^1, respectively, and b and d are in the place values 10^2 and 10^0, respectively. The test says that if $11 \mid ((a + c) - (b + d))$, then $11 \mid abcd$.

Section 5.1

43. Since $11 \times 101010101 = 1,111,111,111$, we know that $11 \mid 1,111,111,111$.
Since $13 \times 8547008547 = 111,111,111,111$, we know that $13 \mid 111,111,111,111$.
Since $17 \times 65359477124183 = 1,111,111,111,111,111$, we know that $17 \mid 1,111,111,111,111,111$.

Section 5.1.

44. (a) Recall that $24! = 24 \times 23 \times 22 \times 21 \times 20 \ldots 6 \times 5 \times 4 \times 3 \times 2 \times 1$. Notice that 3 is a factor of each of the numbers 3, 6, 9, 12, 15, 18, 21, and 24 in the expansion of 24!. If we factor out as many 3's as we

can from each of these numbers, then we can determine the total number of factors of 3 in 24!. We have $3 = 3 \times 1$, $6 = 3 \times 2$, $9 = 3^2 \cdot$, $12 = 3 \times 4$, $15 = 3 \times 5$, $18 = 3^2 \times 2$, $21 = 3 \times 7$, and $24 = 3 \times 8$. Thus, there are 10 factors of 3 in 24!, or we can say that 3^{10} is a factor of 24!. Therefore, $n = 10$ and $3^{10} | 24!$.

(b) We want to find the $n!$ such that the expansion of $n!$ contains 6 factors of 3. We begin by listing counting numbers, noting how many factors of 3 each one contains.

$$1 \times 2 \times \mathbf{3} \times 4 \times 5 \times \mathbf{6} \times 7 \times 8 \times \mathbf{9} \times 10 \times 11 \times \mathbf{12} \times 13 \times 14 \times \mathbf{15}$$
$$\text{1 factor} \qquad \text{1 factor} \quad \text{2 factors} \qquad \text{1 factor} \qquad \text{1 factor}$$

Therefore, in 15! there are 6 factors of 3, so $3^6 | 15!$.

(c) Notice that $12 = 2^2 \times 3$. From the expansion of 24!, find all numbers that contain factors of 2 or 3. Numbers from 1 to 24 that contain factors of 2 or 3 are as follows: 2, 3, 4, 6, 8, 9, 10, 12, 14, 15, 16, 18, 20, 21, 22, and 24. Now factor each of these numbers: 2, 3, 2^2, 2×3, 2^3, 3^2, 2×5, $2^2 \times 3$, 2×7, 3×5, 2^4, 2×3^2, $2^2 \times 5$, 3×7, 2×11, $2^3 \times 3$. There are 22 factors of 2 and 10 factors of 3. Each factor of 12 will contain two 2's and one 3. Since there are only 10 factors of 3, we can only use 20 of the factors of 2 to make 10 factors of 12. Therefore, the largest n is 10, so $12^{10} | 24!$.

Section 5.2

13. Consider the method for finding the number of divisors of any number. Write the number in prime factored form, add 1 to each exponent, and multiply the new exponents. We can create a number with a certain number of divisors by manipulating the exponents when the number is in prime factored form. The <u>smallest</u> number will use the smallest prime factors.

(a) Since 1 is the only number having exactly one divisor, 1 must be the smallest number having exactly one divisor.

(b) To create a number with exactly two divisors, the prime factorization of the number must have one prime factor with an exponent of 1. [$1 + 1 = 2$ divisors.] The smallest prime is 2, and $2^1 = 2$, so 2 is the smallest number with exactly two divisors.

(c) The smallest whole number having three divisors is $2^2 = 4$. Since 3 is prime, the only product that yields 3 is 1×3. Therefore, the exponent on the number in prime factored form must be 2 since $2 + 1 = 3$.

(d) Since 4 can result from 1×4 or 2×2, the small whole numbers having exactly four divisors are $2^3 = 8$ [3 + 1 = 4 divisors] and $2^1 \times 3^1 = 6$ [(1 + 1)(1 + 1) = 4 divisors]. The smallest such number is 6.

(e) Since 5 is prime, the exponent on the prime factor must be 4. Therefore, the smallest whole number with exactly four divisors is $2^4 = 16$ [4 + 1 = 5 divisors].

(f) Since 6 can result from 1×6 or 2×3, small whole numbers having 6 divisors are $2^5 = 32$ [5 + 1 = 6 divisors] and $2^2 \times 3^1 = 12$ [(2 + 1)(1 + 1) = 6 divisors]. The smallest number with 6 divisors is 12.

(g) Since 7 is prime, the exponent on the prime factor must be 6. The smallest whole number with 7 divisors is $2^6 = 64$ [6 + 1 = 7 divisors].

(h) Since 8 can result from 1×8, $2 \times 2 \times 2$, or 2×4, small whole numbers having exactly 8 divisors are $2^7 = 128$ [7 + 1 = 8 divisors], $2^1 \times 3^1 \times 5^1 = 30$ [(1 + 1)(1 + 1)(1 + 1) = 8 divisors], and $2^3 \times 3^1 = 24$ [(3 + 1)(1 + 1) = 8 divisors]. The smallest number with exactly 8 divisors is 24.

Section 5.2

14. (a) The following are examples of numbers with exactly two factors. 2: 1, 2; 3: 1, 3; 5: 1, 5; 7: 1, 7; 11: 1, 11; 13: 1,13. Whole numbers with exactly two factors are prime.

(b) The following are examples of numbers with exactly three factors. 4: 1, 2, 4; 9: 1, 3, 9; 25: 1, 5, 25; 49: 1,7,49; 121: 1,11,121; 169: 1,13,169. Squares of prime numbers have exactly three factors.

(c) The following are examples of numbers with exactly four factors. 6: 1, 2, 3, 6; 8: 1, 2, 4, 8; 10: 1, 2, 5, 10; 14: 1, 2, 7, 14; 15: 1, 3, 5, 10 27: 1, 3, 9, 27. Whole numbers with exactly four factors are perfectcubes or numbers that are the product of 2 primes.

(d) Whole numbers with exactly five factors must be generated by raising prime numbers to the fourth power. By the theorem that provides a method of finding the number of factors of a given number, we would add 1 to the exponent (4 + 1 = 5) to determine the number of factors. Some examples of numbers with exactly five different factors are shown next.

$$16: \ 1, 2, 4, 8, 16$$
$$81: \ 1, 3, 9, 27, 81$$
$$625: \ 1, 5, 25, 125, 625$$
$$2401: \ 1, 7, 49, 343, 2401$$
$$28561: \ 1, 13, 169, 2197, 28561$$
$$14641: \ 1, 11, 121, 1331, 14641$$

Section 5.2

15. $2^{n-1}(2^n - 1)$ is a perfect number when $2^n - 1$ is prime and $n = 1, 2, 3,\dots$ Generate the first four such perfect numbers by substituting $n = 1, 2, 3,\dots$ into the formula and checking to see whether $2^n - 1$ is prime.

n	$2^n - 1$
1	1 not prime
2	3 prime
	perfect number $2^{2-1}(2^2 - 1) = 2(3) = 6$
3	7 prime
	perfect number $2^{3-1}(2^3 - 1) = 4(7) = 28$
4	15 not prime
5	31 prime
	perfect number $2^{5-1}(2^5 - 1) = 16(31) = 496$
6	63 not prime
7	127 prime
	perfect number $2^{7-1}(2^7 - 1) = 64(127) = 8128$

Therefore, the first four such perfect numbers are 6, 28, 496, and 8128.

Section 5.2

16. Since the prime factorization of 24 is $2^3 \times 3$, 24 and any number which is a multiple of 2 or 3 will have a common factor different from 1. All whole numbers from 1 to 24 that are not multiples of 2 or 3 will share at most a common factor of 1 with 24. These numbers are 1, 5, 7, 11, 13, 17, 19, and 23.

Section 5.2

17. Make a table showing how lockers are opened and closed for a small number of students and lockers. The table is shown next, where 0 = open, X = closed. For each locker, read across the table to see how the locker's status changes as students open or close the doors. The final column labeled "Status" indicates the position of each locker door after the tenth student passes through the school.

	Students										
Locker	1	2	3	4	5	6	7	8	9	10	Status
1	0	0	0	0	0	0	0	0	0	0	Open
2	0	X	X	X	X	X	X	X	X	X	Closed
3	0	0	X	X	X	X	X	X	X	X	Closed
4	0	X	X	0	0	0	0	0	0	0	Open
5	0	0	0	0	X	X	X	X	X	X	Closed
6	0	X	0	0	0	X	X	X	X	X	Closed
7	0	0	0	0	0	0	X	X	X	X	Closed
8	0	X	X	0	0	0	0	X	X	X	Closed
9	0	0	X	X	X	X	X	X	0	0	Open
10	0	X	X	X	0	0	0	0	0	X	Closed

Notice that locker numbers 1, 4, and 9 end up open. Since all lockers are closed in the beginning, a locker will be left open only if it has been "changed" an odd number of times. Each locker is "changed" every time the student's number is a factor of the locker number. Recall the theorem which provides a method of finding the number of factors of any number. (The number is written in prime factored form, 1 is added to each exponent, and then the new exponents are multiplied.) For a locker to be "changed" an odd number of times, it must have an odd number of factors. An odd number can only result from the product of odd numbers, By the theorem then, the exponents in the prime factored form of the number would have to all be even since an even number plus 1 is an odd number. Any number in prime factored form that has all even exponents is a perfect square, Therefore, only lockers that have perfect square numbers will remain open.

Section 5.2

18. Since George had no money left over, the price of a candy bar and the price of a can of pop must each be a factor of the total amount of money earned. To find the fewest number of candy bars, we need to find the LCM(15, 48). LCM(15, 48) = LCM $(3 \times 5, 2^4 \times 3) = 2^4 \times 3 \times 5 = 240$. The total amount of money earned was 240 cents. George sold candy bars at 15 cents each. Since $240 = 15 \times 16$, we know that 16 candy bars were sold.

Section 5.2

19. Let c = the price of 1 chicken, d = the price of 1 duck and g = the price of one goose. Since three chickens and one duck sold for as much as two geese, we 'now $3c + d = 2g$. Since one chicken, two ducks and three geese together

sold for $25.00, we know that $c + 2d + 3g = 25$. Now solve for d in the first equation and substitute into the second equation. Solving for d in the first equation yields $d = 2g - 3c$.

Now we substitute $c + 2(2g - 3c) + 3g = 25$

Simplify $c + 4g - 6c + 3g = 25$

$$7g - 5c = 25$$
$$7g = 25 + 5c$$
$$7g = 5(5 + c)$$

Since the exact dollar amount is asked for, only whole numbers need to be considered. Notice that $7g$ must be a multiple of 5 since $7g = 5(5 + c)$. If $g = \$5$ then $7(5) - 5c = 25$ and $c = \$2$. Therefore, $d = 2(5) - 3(2) = 10 - 6 = \$4$. The prices of one chicken, one duck, and one goose are $2, $4, and $5, respectively.

Section 5.2

20.   Write each number in the set in prime factored form. $\{10, 20, 40, 80, 160, \ldots\} = \{2 \times 5, 2^2 \times 5, 2^3 \times 5, 2^4 \times 5, 2^5 \times 5, \ldots\}$ Any perfect square can be written as a product of some number and itself. This means that if the perfect square were written in prime factored form, each prime factor would have an even exponent. Since each number in the set has only one factor of 5, none of them can be perfect squares.

Section 5.2

21.   Since each of the three digits in the number must be prime, it is only necessary to consider digits 2, 3, 5, and 7. We seek the largest prime number, so make a list of possible three-digit numbers using the largest prime digits first. Stop when the first three-digit prime number has been found.

Three Digit Number	Prime or Composite
777	Composite $777 = 7 \times 111$
775	Composite $775 = 5 \times 155$
773	Prime

Section 5.2

22.   (a) Since we want to find a divisor greater than 1 for every such number, we use variables. Let *abba* be any four-digit palindrome. Then *baab* is a new palindrome formed by interchanging unlike digits. Add these two numbers using expanded form, as shown next.

$$abba + baab = 1000a + 100b + 10b + a + 1000b +$$
$$100a + 10a + b$$
$$= 1111a + 1111b$$
$$= 1111(a + b).$$

Since $1111 = 11 \times 101$, we know that 11, 101, and 1111 all divide the sum.

(b) 1111 is the largest whole number that divides the sum.

Section 5.2
23. In an additive magic square, all rows, columns, and diagonals must add to the same number. Adding row 1 yields a total of 120. Row 2 contains one empty spot. The missing prime must be $120 - 43 - 31 - 5 = 41$. Now column 4 contains one empty spot. The missing prime in column 4 is $120 - 37 - 41 - 29 = 13$. The missing prime in column 3 is $120 - 19 - 5 - 23 = 73$. Consider row 3. There are two missing prime numbers. Their total must be $120 - 73 - 29 = 18$. The only primes that have not been used and that add to 18 are 11 and 7. Placing 7 at the intersection of row 3 and column 1 forces the missing prime in column 1 to be $120 - 3 - 43 - 7 = 67$. Placing 11 at the intersection of row 3 and column 2 forces the missing prime in column 2 to be $120 - 61 - 31 - 11 = 17$. The numbers in the magic square are:

3	61	19	37
43	31	5	41
7	11	73	29
67	17	23	13

Section 5.2
24. If the number of cards minus 1 is divisible by 2, 3, and 5, then the fewest number of cards that could satisfy the conditions is 1 more than the LCM of 2, 3, and 5. $LCM(2, 3, 5) = 2 \times 3 \times 5 = 30$. Thus, the fewest number of cards possible is $30 + 1 = 31$.

Section 5.2
25. $343 = 7 \times 49$, so 343 is divisible by 7. Consider any three-digit number of this type in expanded form: $100a + 10b + a$. If we assume $a + b = 7$ then $b = 7 - a$. Substituting $7 - a$ for b in the three-digit number, we have
$$100a + 10(7 - a) + a = 100a + 70 - 10a + a$$
$$= 91a + 70$$
$$= 7(13a + 10).$$
Therefore, since 7 is a factor, the number is divisible by 7.

Section 5.2

26. Write each number in the set in prime factored form.
$\{18, 96, 54, 27, 42\} = \{2 \times 3^2, 2^5 \times 3, 2 \times 3^3, 3^3, 2 \times 3 \times 7\}$.
To find the pair with the greatest GCF, consider the GCF for all possible pairs.
GCF(18, 96) = $2 \times 3 = 6$
GCF(18, 54) = $2 \times 3^2 = 18$
GCF(18, 27) = $3^2 = 9$
GCF(18, 42) = $2 \times 3 = 6$
GCF(96, 54) = $2 \times 3 = 6$
GCF(96, 27) = 3
GCF(96, 42) = $2 \times 3 = 6$
GCF(54, 27) = 3^3 = 27
GCF(54, 42) = $2 \times 3 = 6$
GCF(27, 42) = 3
The pair with the greatest GCF is 54 and 27.
GCF(54, 27) = 27. To find the pair with the smallest LCM, consider the LCM for all possible pairs.
LCM(18, 96) = $2^5 \times 3^2 = 288$
LCM(18, 54) = 2×3^3 = 54
LCM(18, 27) = 2×3^3 = 54
LCM(18, 42) = $2 \times 3^2 \times 7 = 126$
LCM(96, 54) = $2^5 \times 3^3 = 864$
LCM(96, 27) = $2^5 \times 3^3 = 864$
LCM(96, 42) = $2^5 \times 3 \times 7 = 672$
LCM(54, 27) = 2×3^3 = 54
LCM(54, 42) = $2 \times 3^3 \times 7 = 378$
LCM(27, 42) = $2 \times 3^3 \times 7 = 378$
Three pairs have a least common multiple (LCM) of 54, which is the smallest: 18 and 54, 18 and 27, 54 and 27.

SOLUTIONS - PART A PROBLEMS

Chapter 6: Fractions

Section 6.1
19. (a) $1/4 = 25/100 = 25/10^2$
    (b)   Does not work
    (c) $1/5 = 2/10$
    (d) $3/25 = 12/100 = 12/10^2$
    (e) $1/8 = 125/1000 = 125/10^3$
    (f) $55/80 = 6875/10000 = 6875/10^4$
    (g)   Does not work
    (h) $9/40 = 225/1000 = 225/10^3$
    (i)   Does not work
    (j) $14/70 = 2/10 = 2/10^1$
    (k)   Does not work
    (l) $57/64 = 890,625/1,000,000 = 890,625/10^6$
    (m) $27/60 = 9/20 = 45/100 = 45/10^2$
    (n) $75/128 = 5,859,375/10,000,000 = 5,859,375/10^7$
    (o)   Does not work
    (p) $52/130 = 4/10 = 4/10^1$

All fractions that do not work have, when written in simplest form, denominators that do not divide a power of 10. In order for a denominator to divide a power of 10, its prime factors can be only 2's and/or 5's. If a denominator has other prime factors, then it will never have a power of 10 as a multiple.

Factor the numerators and denominators for those fractions that work:

$1/2^2$, $1/5$, $3/5^2$, $1/2^3$, $(5 \times 11)/(2^4 \times 5)$, $3^2/(2^3 \times 5)$, $(2 \times 7)/(2 \times 5 \times 7)$, $(3 \times 19)/2^6$, $3^3/(2^2 \times 3 \times 5)$, $(3 \times 5^2)/2^7$, $(2^2 \times 13)/(2 \times 5 \times 13)$.

If we reduced each of these fractions, then the fraction in simplest form would only have 2's and/or 5's in the denominator.

Section 6.1
20. Since x, y, and z are counting numbers, we know none of them is zero.
    (a) If the denominators are the same for two fractions, then in order for them to be equal, the numerators must be the same. That is, if $\dfrac{x}{z} = \dfrac{y}{z}$, then $x = y$.

(b) If the numerators are the same for two fractions, then in order for them to be equal, the denominators must be the same. That is, if $\dfrac{x}{y} = \dfrac{x}{z}$, then $y = z$.

Section 6.1

21. (a) There are many ways to divide the hexagon into two equal pieces. Notice that if it is folded down the center vertically, then the halves match exactly.

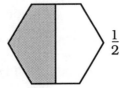

(b) Construct three segments extending from the center perpendicular to every other side.

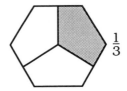

(c) Consider our picture from part (a). We can divide each half into two equal parts by constructing a horizontal segment through the center of the hexagon. This will divide the hexagon into four equal pieces.

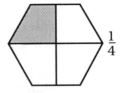

(d) Consider our picture from part (b). We can divide each third into two equal parts by constructing three more segments from the center perpendicular to the three remaining sides. This will divide the hexagon into six equal parts as shown.

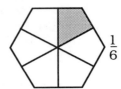

Alternatively, consider the hexagon which has been divided into thirds in the problem statement. Each third can be divided into two equal parts by constructing three more segments from the center to the three remaining vertices. This will divide the hexagon into six equal parts.

$\frac{1}{6}$

(e) Consider the first figure from part (d). Each sixth can be divided into two equal pieces by constructing six segments from the center to each vertex. This divides the hexagon into 12 equal pieces.

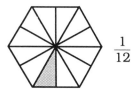

$\frac{1}{12}$

(f) Consider the second figure from part (d). The hexagon is divided into six equal pieces. Consider one triangular piece. We can divide it into four equal pieces by connecting consecutive midpoints for the sides, as shown next.

Doing this for each of the six triangles in the hexagon will divide it into 24 equal pieces.

$\frac{1}{24}$

Section 6.1
22. Consider several examples:

$$\frac{286}{583} = \frac{11 \times 26}{11 \times 53}$$

$$\frac{2886}{5883} = \frac{111 \times 26}{111 \times 53}$$

$$\frac{28886}{58883} = \frac{1111 \times 26}{1111 \times 53}$$

$$\frac{288886}{588883} = \frac{11111 \times 26}{11111 \times 53}$$

Notice that the number common to the numerator and denominator in each case has one more 1 than the number of 8's in the numerator of the original fraction.

$$\frac{288...86}{588...83} = \frac{26 \times 111...1}{53 \times 111...1} = \frac{26}{53}$$

Section 6.1

23. (a) False. Compare 23/100 to 230/10000. In this case the numerator increased, but the fraction decreased because the denominator was increased by a larger power of 10. 23/100 > 230/10000.
    (b) False. Compare 77/100 to 7777/1000. In this case, the denominator increased, but the fraction increased because the numerator was also significantly increased. 77/100 < 7777/1000.
    (c) True. Compare fractions with the same denominator: 5/17 < 9/17 < 13/17. If the denominator remains the same and the numerator increases, then the fraction increases.
    (d) True. Compare fractions with the same numerator: 3/10 < 3/20 < 3/30. If the numerator remains the same and the denominator increases, then the fraction decreases.

Section 6.1

24. The sum of the numerator and denominator is a one-digit perfect square. The only one-digit squares are 1, 4, and 9. Consider all proper fractions whose numerators and denominators add to 1, 4, or 9. From this list, find the fraction such that the product of its numerator and denominator is a cube.

Fractions	Sum	Product
1/3	4	3
1/8	**9**	**8 = 2^3**
2/7	9	14
3/6	9	18
4/5	9	20

$\frac{1}{8}$ is the desired fraction.

Section 6.2

15. (a) If $a/b < c/d$ and $c/d < e/f$, then there are nonzero m/n and p/q so that $a/b + m/n = c/d$, and $c/d + p/q = e/f$. Substituting $c/d = a/b + m/n$ into the second equation, we have $(a/b + m/n) + p/q = e/f$. So by the associative property, we have $a/b + (m/n + p/q) = e/f$ where $m/n + p/q$ is a nonzero fraction. Therefore, $a/b < e/f$, by the alternative definition of less than.

(b) If $a/b < c/d$, then there is a nonzero m/n so that $a/b + m/n = c/d$. Then we have

$$\begin{aligned} c/d + e/f &= (a/b + m/n) + e/f && \text{by substitution} \\ &= m/n + (a/b + e/f) && \text{by commutativity} \\ & && \text{and} \\ & && \text{associativity for} \\ & && \text{fraction addition.} \end{aligned}$$

Therefore, $a/b + e/f < c/d + e/f$ by the alternative definition of less than.

(c) If $a/b < c/d$, then there is a nonzero m/n so that $a/b + m/n = c/d$. Then we have

$$\begin{aligned} c/d \times e/f &= (a/b + m/n) \times e/f && \text{by substitution} \\ &= a/b \times e/f + m/n \times e/f && \text{by the distributive} \\ & && \text{property.} \end{aligned}$$

Since $m/n \times e/f$ is a nonzero fraction, we know that $a/b \times e/f < c/d \times e/f$ by the alternative definition of less than.

Section 6.2

16. Consider a simpler problem with fewer terms. Make a list and find a pattern.

Terms	Sum
1	$1/2$
2	$1/2 + (1/2)^2 = 1/2 + 1/4 = 3/4$
3	$1/2 + (1/2)^2 + (1/2)^3 = 1/2 + 1/4 + 1/8 = 7/8$
4	$1/2 + (1/2)^2 + (1/2)^3 + (1/2)^4 = 1/2 + 1/4 + 1/8 + 1/16 = 15/16$
5	$1/2 + (1/2)^2 + (1/2)^3 + (1/2)^4 + (1/2)^5 = 1/2 + 1/4 + 1/8 + 1/16 + 1/32 = 31/32$

Notice that there is a relationship between the number of terms and the numerator and denominator. For n terms, the numerator is $2^n - 1$ and the denominator is 2^n. For the sum $1/2 + 1/2^2 + \ldots + 1/2^{100}$, the numerator is $2^{100} - 1$ and the denominator is 2^{100}. Therefore, the sum is $(2^{100} - 1)/2^{100}$.

Section 6.2

17. Sally, her brother, and a third person each own a fraction of one whole restaurant. Together, Sally and her brother own $1/3 + 1/4 = 4/12 + 3/12 = 7/12$ of the restaurant. The third person owns the rest, or $1 - 7/12 = 12/12 - 7/12 = 5/12$. Therefore, the third person owns $5/12$ of the restaurant.

Section 6.2

18. Let t = John's age at death or the number of years John lived. John's life can be broken down into non-overlapping time periods, as shown next.

    Period

    Childhood: one-quarter of his life $= \frac{1}{4}t$

    College: one-sixth of his life $= \frac{1}{6}t$

    Teaching: one-half of his life $= \frac{1}{2}t$

    Retirement: last six years $= 6$

    If all of these periods are added, the total is t years.

    $$\frac{1}{4}t + \frac{1}{6}t + \frac{1}{2}t + 6 = t$$
    $$\frac{3}{12}t + \frac{2}{12}t + \frac{6}{12}t + 6 = t$$
    $$\frac{11}{12}t + 6 = t$$
    $$11t + 72 = 12t$$
    $$72 = t.$$

    Therefore, John was 72 years old when he died.

Section 6.2

19. (a) $3/6 = 1/2$
    (b) 18 hits this season
    (c) 56 at-bats this season
    (d) $18/56 = 9/28$
    (e) Compare $15/50 + 3/6$ and $15/50 + 1/2$ using "Baseball Addition"
    $15/50 + 3/6 = (15 + 3)/(50 + 6) = 18/56 = 9/28$
    $15/50 + 1/2 = (15 + 1)/(50 + 2) = 16/52 = 4/13$
    When fractions are replaced by equivalent fractions, the answers are not the same.

Section 6.2

20. (a) $1/2 + 1/3 + 1/6 = 3/6 + 2/6 + 1/6 = 6/6 = 1$
    (b) $1/2 + 1/4 + 1/7 + 1/14 + 1/28 = 14/28 + 7/28 + 4/28 + 2/28 + 1/28 = 28/28 = 1$

(c) This result is true for 496. Consider the factors of 496: 1, 2, 4, 8, 16, 31, 62, 124, 248, and 496. Adding the corresponding fractions, we have $1/2 + 1/4 + 1/8 + 1/16 + 1/31 + 1/62 + 1/124 + 1/248 + 1/496 = 248/496 + 124/496 + 62/496 + 31/496 + 16/496 + 8/496 + 4/496 + 2/496 + 1/496 = 1$. Notice that $1 + 2 + 4 + 8 + 16 + 31 + 62 + 124 + 248 = 496$. A number whose proper factors have a sum equal to the number is called a perfect number. Thus, 496 is a perfect number. The result described in parts (a) - (c) is true for all perfect numbers.

Section 6.2

21. David is subtracting as if he is using base 10 numbers. When he borrowed, he thought he borrowed 10 rather than 5. If David could use base 5 blocks, he could see that when he borrows 1 long, it is equal to 5/5.

Section 6.2

22. It is possible to add unitary fractions with different odd denominators to obtain 1. Find the least common denominator for the given sum: LCD = $3^3 \times 5 \times 7 = 945$. Now adding the unitary fractions, we have $1/3 + 1/5 + 1/7 + 1/9 + 1/15 + 1/21 + 1/27 + 1/35 + 1/63 + 1/105 + 1/135 = 315/945 + 189/945 + 135/945 + 105/945 + 63/945 + 45/945 + 35/945 + 27/945 + 15/945 + 9/945 + 7/945 = 945/945 = 1$.

Section 6.2

23. In the example, $1/2 = 1/3 + 1/6$. Notice that the least common multiple of 2 and 3 is 6. The unitary fractions are related as follows: $1/n = 1/(n + 1) + 1/(n(n + 1))$.
    (a) $1/5 = 1/6 + 1/30$.
    (b) $1/7 = 1/8 + 1/56$.
    (c) $1/17 = 1/18 + 1/306$.

Section 6.2

24. $$\frac{1+3}{5+7} = \frac{4}{12} = \frac{1}{3}$$
    $$\frac{1+3+5}{7+9+11} = \frac{9}{27} = \frac{1}{3}$$
    $$\frac{1+3+5+7}{9+11+13+15} = \frac{16}{48} = \frac{1}{3}$$

Two other such fractions are:
    $$\frac{1+3+5+7+9}{11+13+15+17+19} = \frac{25}{75} = \frac{1}{3}$$
    $$\frac{1+3+5+7+9+11}{13+15+17+19+21+23} = \frac{36}{108} = \frac{1}{3}$$

Notice from the hint, the sum of the first n odd numbers is n^2. If there are n terms in the numerator, we can write the numerator as $1+3+5+\cdots+(2(n)-1)$. Since the denominator is the sum of the next n odd numbers, we can write it as $(2n+1)+(2n+3)+\cdots+(2(2n)-1)$. Notice that if the denominator had been the sum $1+3+5+\cdots+(2(2n)-1)$, then the sum would have been $(2n)^2 = 4n^2$. But remember, the first n terms of this sum are in the numerator, and we know their sum is n^2. Therefore, the sum of the terms in the denominator is $4n^2 - n^2 = 3n^2$. Thus, the fraction can be written

$$\frac{n^2}{3n^2} = \frac{1}{3}.$$

Section 6.2

25. Consider a simpler problem. Suppose there are only four players. Visualize this by drawing a square. The four players are represented by the four vertices of the square. The number of matches would be the same as the number of line segments required to connect each vertex with each of the other vertices.

Player A plays D, C, and B. (3 games)

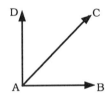

Additionally, player B plays D and C. (2 more games) Notice that player B already played A.

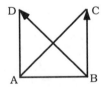

Finally, player C plays D. (1 more game)
Notice player C already played B and A.
Player D has played everyone at this point.
Total number of matches: $3 + 2 + 1 = 6$ matches.

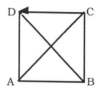

If the tournament begins with eight players, we can visualize this by drawing a regular octagon. Vertex A can be connected to the 7 other vertices. (7 games).

Vertex B connects to 6 additional vertices. Vertex C connects to 5 additional vertices; vertex D, 4 additional vertices; vertex E, 3 additional vertices; vertex F, 2 additional vertices; vertex G, 1 additional vertex. The total number of matches is: $7 + 6 + 5 + 4 + 3 + 2 + 1 = 28$.

Section 6.3

21. Suppose that each load required 2 cups of detergent. In that case, to find the number of loads that could be done, divide the number of cups in the box by 2. Since in the problem there are 40 cups in the box, and each load requires $1\frac{1}{4}$ cups, divide the total number of cups in the box by $1\frac{1}{4}$: $40 \div 1\frac{1}{4} = 40 \div \frac{5}{4} = 40 \times \frac{4}{5} = 32$. Therefore, you can do 32 loads.

Section 6.3

22. Let x = the number of barrels of oil used for transportation per day in the United States. Then $\frac{5}{7}$ of the total barrels used per day for transportation is equal to the total number of barrels produced, So,

$$\frac{5}{7}x = 7,630,000$$
$$5x = 7(7,630,000)$$
$$5x = 53,410,000$$
$$x = 10,682,000.$$

Therefore, approximately 10,700,000 barrels of oil are used each day for transportation in the United States.

Section 6.3

23. Let x = the total number of students enrolled. Since all but $\frac{1}{16}$ of the students enrolled participated, we know that $\frac{15}{16}$ of the students enrolled in the school participated in the activities. So, we have

$$\frac{15}{16}x = 405$$
$$15x = 16(405)$$
$$15x = 6480$$
$$x = 432.$$

Therefore, 432 students attend the school.

Section 6.3

24. (a) Try a simpler problem. Suppose that the directions recommended mixing 2 ounces of the concentrate with 1 gallon of the water. To find out how many gallons of mixture can be made from the bottle of concentrate, we need to figure out how many 2-ounce portions there are in the 32-ounce bottle of concentrate.

Divide: $\dfrac{32 \text{ ounces}}{2 \text{ ounces per portion}}$ = 16 portions

Therefore, since there are 16 portions, and each portion mixes with 1 gallon of water, we see that approximately 16 gallons of mixture can be made. For portions of $2\frac{1}{2}$ ounces of concentrate, divide again to find out how many gallons of mixture can be made.

$$\dfrac{32 \text{ ounces}}{2\frac{1}{2} \text{ ounces per portion}} = 12\frac{4}{5} = 12.8 \text{ portions.}$$

Therefore, approximately 13 gallons of mixture can be made.

(b) For portions of $1\frac{3}{4}$ ounces of concentrate, we can make

$$\dfrac{32 \text{ ounces}}{1\frac{3}{4} \text{ ounces per portion}} = 18\frac{2}{7} \approx 18.3 \text{ portions.}$$

Therefore approximately 18.3 gallons can be made or 18.3 − 12.8 = 5.5 more gallons.

Section 6.3

25. Since Kathleen used $\frac{2}{5}$ of her time to study math, $\frac{3}{20}$ to study Spanish, and $\frac{1}{3}$ to study biology, she spent $\frac{2}{5} + \frac{3}{20} + \frac{1}{3} = \frac{53}{60}$ of her time studying subjects other than English. Since the remaining 35 minutes of her time was devoted to studying English, we know that $1 - \frac{53}{60} = \frac{7}{60}$ of her time was devoted to English.

If x is her total study time, then we have

$$\frac{7}{60}x = 35$$
$$7x = 35(60)$$
$$7x = 2100$$
$$x = 300.$$

Kathleen spent 300 minutes, or 5 hours, studying in one evening! (She must be an A student.) Since $\frac{3}{20}$ of her time was spent studying Spanish, she spent $\frac{3}{20}(300) = 45$ minutes studying Spanish. Notice that she spent $\frac{2}{5}(300) = 120$ minutes = 2 hours, studying math, ... a good example to follow.

Section 6.3

26. Let x = the number of employees originally enrolled in a fitness program. By March 2nd, four-fifths of them or $\frac{4}{5}x$ were still participating. By May 2nd, $\frac{5}{6}$ of those who were participating on March 2nd were still participating, that is $\frac{5}{6}\left(\frac{4}{5}x\right) = \frac{2}{3}x$. By July 2nd, $\frac{9}{10}$ of those who were participating on May 2nd were still participating, or $\frac{9}{10}\left(\frac{2}{3}x\right) = \frac{3}{5}x$. Since 36 of the original participants were still active on July 2nd, we know that

$$\frac{3}{5}x = 36$$
$$3x = 5(36)$$
$$3x = 180$$
$$x = 60.$$

Therefore, 60 employees were originally enrolled.

Section 6.3

27. Tammy passes the grocery store at 7:40 and the bicycle shop at 7:45. In 5 minutes she walks 1/2 − 1/3 = 1/6 of the distance to school. At 7:45 when she passes the bicycle shop, she still has 1/2 the distance to go. Since 1/2 = (1/6) × 3, it takes 5 × 3 = 15 minutes for the second half of the trip. Tammy arrives at school at 8:00 a.m.

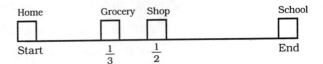

Section 6.3

28. (a) If the recipe is doubled, then the amount of flour required is doubled. The recipe calls for $1\frac{1}{4}$ cups of flour. Double the recipe calls for twice as much flour.

$$2 \times 1\frac{1}{4} = 2 \times \frac{5}{4} = \frac{10}{4} = \frac{5}{2} = 2\frac{1}{2} \text{ cups of flour.}$$

(b) If the recipe is cut in half, then the amount of flour in the recipe is reduced by 1/2 of the original amount.

$$\frac{1}{2} \times 1\frac{1}{4} = \frac{1}{2} \times \frac{5}{4} = \frac{5}{8} \text{ cups of flour.}$$

(c) Notice that one recipe makes 3 dozen cookies. One recipe = 3 dozen, so 1/3 recipe = 1 dozen. To make 5 dozen cookies, then, you need 5/3 of a recipe since 5(1 dozen) = 5(1/3 recipe) = 5/3 recipe. If you want to make 5/3 of a recipe, then the amount of flour required for one recipe should be multiplied by 5/3.

$$\frac{5}{3} \times 1\frac{1}{4} = \frac{5}{3} \times \frac{5}{4} = \frac{25}{12} = 2\frac{1}{12} \text{ cups of flour.}$$

Section 6. 3

29. Let x = the total number of games played. Since Gale started in $\frac{3}{8}$ of the games played, she started in $\frac{3}{8}x$ games. Since Sandy started in 1 more game than Gale, she started in $\frac{3}{8}x + 1$ games. Since Ruth started in half as many games as Sandy, she started in $\dfrac{\frac{3}{8}x + 1}{2} = \dfrac{\frac{3x+8}{8}}{2} = \dfrac{3x+8}{16}$ games. These pitchers were the only three

starters, so the total number of games, x, must equal the sum of the number of games each pitcher started.

$$x = \frac{3}{8}x + \frac{3}{8}x + 1 + \frac{3x+8}{16}$$
$$16x = 6x + 6x + 16 + 3x + 8$$
$$16x = 15x + 24$$
$$x = 24 \text{ games.}$$

Therefore, Gale started $\frac{3}{8}$ (24) = 9 games, Sandy started $9 + 1 = 10$ games, and Ruth started $\frac{10}{2}$ = 5 games.

Section 6.3

30. Each year the value of the equipment is $\frac{1}{20}$ less than its value the previous year. Thus, each year it retains $1 - \frac{1}{20}$ = $\frac{19}{20}$ of its value from the previous year.

(a) The office equipment was purchased for $60,000. After one year its value is $\frac{19}{20}$ (60,000) = $57,000. After the second year, the value is $\frac{19}{20}$ (57,000) = $54,150.

(b) Continue calculating the values after successive years.

Year	Value (in dollars)
1	$\frac{19}{20}$ (60,000) = 57,000
2	$\frac{19}{20}$ (57,000) = 54,150
3	$\frac{19}{20}$ (54,150) = 51,442.50
4	$\frac{19}{20}$ (51,442.50) = 48,870.38
5	$\frac{19}{20}$ (48,870.38) = 46,426.86
6	$\frac{19}{20}$ (46,426.86) = 44,105.51
7	$\frac{19}{20}$ (44,105.51) = 41,900.24
8	$\frac{19}{20}$ (41,900.24) = 39,805.23

At the end of the 8th year, the piece of equipment will have a value less than $40,000.

Section 6.3

31. Let n and m be the two counting numbers. Set up two equations from the information given. Consider n. One more than the number itself is $n + 1$. So $\dfrac{n}{n + 1} = \dfrac{1}{5}$. Now consider m. One-fifth of this number itself is $\dfrac{1}{5}m = \dfrac{m}{5}$. So $\dfrac{m}{m + 1} = \dfrac{m}{5}$. We want the product of the two numbers, n and m, so we will solve for n and m in the equations. Consider that $\dfrac{n}{n + 1} = \dfrac{1}{5}$. Cross multiplying gives $5n = n + 1$. So $4n = 1$ and $n = \dfrac{1}{4}$. Now consider that $\dfrac{m}{m + 1} = \dfrac{m}{5}$. Since these two fractions are equal and their numerators are the same, their denominators must also be equal. Thus, $m + 1 = 5$, so $m = 4$. Therefore, the product of the two numbers is $m \times n = 4 \times \dfrac{1}{4} = 1$.

Section 6.3

32. (a) Multiply the denominator by 5: $\dfrac{11}{16} \div 5 = \dfrac{11}{16 \times 5} = \dfrac{11}{80}$.

    (b) Yes. $\dfrac{a}{b} \div n = \dfrac{a}{b} \div \dfrac{n}{1} = \dfrac{a}{b} \times \dfrac{1}{n} = \dfrac{a}{b \times n}$.

    (c) Consider $10\dfrac{9}{16} \div 2$. Since dividing by 2 is the same as multiplying by $\dfrac{1}{2}$, we can calculate this quotient quickly by dividing 10 in half and multiplying the denominator of the fraction by 2. $10\dfrac{9}{16} \div 2 = 5\dfrac{9}{32}$.

    Now consider $5\dfrac{3}{8} \div 2$. Since 5 is not an even number, we can make this problem simpler by borrowing 1 from 5. Then we divide the whole number in half and multiply the denominator of the fraction by 2. We have $5\dfrac{3}{8} = 4 + 1\dfrac{3}{8} = 4\dfrac{11}{8}$, so $5\dfrac{3}{8} \div 2 = 4\dfrac{11}{8} \div 2 = 2\dfrac{11}{16}$.

Section 6.3

33. (a) Sam finds the least common denominator first. When he multiplies the fractions, he forgets to multiply the denominators. Sam would write $\dfrac{3}{4} \times \dfrac{1}{6} = \dfrac{9}{12} \times \dfrac{2}{12} = \dfrac{18}{12} = \dfrac{3}{2} = 1\dfrac{1}{2}$. Sandy inverts the second fraction before

multiplying. Sandy would write $\dfrac{5}{6} \times \dfrac{3}{8} = \dfrac{5}{6} \times \dfrac{8}{3} = \dfrac{40}{18} = \dfrac{20}{9} = 2\dfrac{2}{9}$.

(b) Sam is confusing multiplication with addition. Sandy is confusing multiplication with division.

Section 6.3

34. Notice that it may be necessary to use equivalent fractions in the equation. There is more than one solution in each case.

   (a) $\dfrac{8}{15}$ can be thought of as $\dfrac{8}{5} \times \dfrac{1}{3}$. However, the number 1 cannot be used, so the fraction $\dfrac{1}{3}$ must be expressed as an equivalent fraction. For example, we know that $\dfrac{1}{3} = \dfrac{3}{9}$. Therefore, $\dfrac{8}{5} \times \dfrac{3}{9} = \dfrac{8}{15}$.

   (b) Recall that when dividing fractions, we invert the second one and multiply. Notice that $\dfrac{5}{4}$ is the same as $\dfrac{5}{8} \times 2$, which is the same as $\dfrac{5}{8} \div \dfrac{1}{2}$. Since we cannot use 1 or 2, we create an equivalent fraction. For example, $\dfrac{1}{2} = \dfrac{3}{6}$. Therefore, $\dfrac{5}{8} \div \dfrac{3}{6} = \dfrac{5}{4}$.

   (c) $\dfrac{6}{5}$ can be thought as $\dfrac{3}{5} \times 2$. There are many ways to create 2 using equivalent fractions. For example, $2 = \dfrac{8}{4}$, so $\dfrac{3}{5} \times \dfrac{8}{4} = \dfrac{6}{5}$.

   (d) $\dfrac{16}{5}$ can be thought as $\dfrac{8}{5} \times 2$, which is the same as $\dfrac{8}{5} \div \dfrac{1}{2}$. Create $\dfrac{1}{2}$ using an equivalent fraction. For example, $\dfrac{1}{2} = \dfrac{3}{6}$, so $\dfrac{8}{5} \div \dfrac{3}{6} = \dfrac{16}{5}$.

Section 6.3

35. Let x = the number of apples in the store. Then we have $\dfrac{1}{4}x + \dfrac{1}{5}x + \dfrac{1}{6}x = 37$. The least common multiple of 4, 5, and 6 is 60. So we can write

$$\dfrac{15x}{60} + \dfrac{12x}{60} + \dfrac{10x}{60} = 37$$

$$\dfrac{37x}{60} = 37$$

$$x = 60.$$

There were 60 apples in the store.

Section 6.3

36. If there are 5 elected officers, then let them be represented by A, B, C, D, and E. If only two of the officers will attend a state-wide meeting, then make a list of all possible pairs of delegates.

AB	BC	CD	DE	
AC	BD	CE		10 pairs in all
AD	BE			
AE				

Notice that AB and BA are the same pair of delegates. We do not need to count such pairs twice. Therefore, there are 10 ways that two delegates can be selected from the five officers.

SOLUTIONS - PART A PROBLEMS

Chapter 7: Decimals, Ratio, Proportion, and Percent

Section 7.1

16. To be an additive magic square, the three rows, the three columns, and the two diagonals must add up to the same number. The sum of all the numbers is 10.48 + 15.72 + ... + 47.16 + 52.4 = 282.96. Since, for example, each of the three rows must add up to the same number, each row must add up to 282.96/3 = 94.32. Guess and test to find sets of three numbers that add up to 94.32. The solution is shown below.

26.2	20.96	47.16
52.4	31.44	10.48
15.72	41.92	36.68

Section 7.1

17. Let x = the amount of money Gary had before cashing the check. Gary bought the following:

Items	Cost
2 magazines	2($1.95) = $3.90
1 book	$5.95
1 record	$5.98
Total cost	$15.83

After cashing the check, Gary had $x + 29.35$ dollars. He spent $15.83 and had $21.45 left.
$$x + 29.35 - 15.83 = 21.45$$
$$x + 13.52 = 21.45$$
$$x = 7.93.$$
Gary had $7.93 before cashing the check.

Section 7.1

18. Let x = original value of the car. After one year, the car is worth $\frac{4}{5}x$. After two years have passed, the car is worth $\frac{4}{5}\left(\frac{4}{5}x\right) = \frac{16}{25}x$. After three years, the car is worth $\frac{4}{5}\left(\frac{16}{25}x\right) = \frac{64}{125}x$.

If the car is worth $16,000 after 3 years have passed, then we know that

$$\$16{,}000 = \frac{64}{125}x$$
$$16000 \times \frac{125}{64} = x$$
$$\$31{,}250 = x.$$

The original value of the car was $31,250.

Section 7.1

19. Let x = the number of candy bars Todd ate on the first day. On the second day, Todd ate $x + 2$ candy bars. On the third, fourth, and fifth days, Todd ate $x + 4$, $x + 6$, and $x + 8$ candy bars, respectively. Since he ate as many candy bars during the last 2 days as he did during the first 3 days,

$$x + (x + 2) + (x + 4) = (x + 6) + (x + 8)$$
$$3x + 6 = 2x + 14$$
$$x = 8.$$

Therefore, Todd ate 8 candy bars on the first day, 10 on the second day, 12 on the third day, 14 on the fourth day, and 16 on the fifth day.

Section 7.2

14. Recall that Rate × Time = Distance. Since we know the distance in meters and the rate in meters per light year, we can find the time in light years.

$$T = \frac{D}{R} = \frac{1.5 \times 10^{19} \text{ meters}}{9.46 \times 10^{15} \dfrac{\text{meters}}{\text{light year}}} \approx 1585.6 \text{ light years.}$$

Therefore, the distance from Earth to Deneb is approximately 1600 light years.

Section 7.2

15. The decimal expansion is nonterminating. If the fraction a/b is in simplest form, then a/b has a repeating decimal that does not terminate if, and only if, b has a prime factor other than 2 or 5. Since 151 is prime, the fraction is in simplest form, and the denominator does not contain 2 or 5 as a factor (it does not end in 0 or 5 and is not even), the fraction is nonterminating.

Section 7.2

16. For the decimal expansion of a fraction to terminate, the prime factors of the denominator must be only 2's and/or 5's. Make a list of such fractions and find a relationship between the fraction and the number of places in the terminating decimal.

Fraction	Decimal Expansion
$1/2 = 1/2^1$	0.5
$1/4 = 1/2^2$	0.25
$1/8 = 1/2^3$	0.125
$1/16 = 1/2^4$	0.0625

Notice that the number of places in the decimal expansion is the same as the exponent in the denominator of the fraction. So we can now write examples of the desired fractions. (Note: Other answers are possible.)

(a) $1/8 = 0.125$

(b) $1/16 = 0.0625$

(c) $1/256 = 1/2^8 = 0.00390625$

(d) $1/2^{17}$

Section 7.2

17. When mentally converting decimals into fractions, $1/9$ can be substituted for the decimal $0.\overline{1}$ since they are equal.

(a) Since $0.\overline{3} = 3(0.\overline{1})$, $0.\overline{3} = 3(1/9) = 3/9 = 1/3$.

(b) Since $0.\overline{5} = 5(0.\overline{1})$, $0.\overline{5} = 5(1/9) = 5/9$.

(c) Since $0.\overline{7} = 7(0.\overline{1})$, $0.\overline{7} = 7(1/9) = 7/9$.

(d) Since $2.\overline{8} = 2 + 8(0.\overline{1})$, $2.\overline{8} = 2 + 8(1/9) = 2\frac{8}{9}$.

(e) Since $5.\overline{9} = 5 + 9(0.\overline{1})$, $5.\overline{9} = 5 + 9(1/9) = 5 + 1 = 6$.

Section 7.2

18. When mentally converting decimals into fractions, $1/99$ can be substituted for the decimal $0.\overline{01}$ since they are equal.

(a) $0.\overline{03} = 3(0.\overline{01}) = 3(1/99) = 3/99 = 1/33$.

(b) $0.\overline{05} = 5(0.\overline{01}) = 5(1/99) = 5/99$.

(c) $0.\overline{07} = 7(0.\overline{01}) = 7(1/99) = 7/99$.

(d) $0.\overline{37} = 37(0.\overline{01}) = 37(1/99) = 37/99$.

(e) $0.\overline{64} = 64(0.\overline{01}) = 64(1/99) = 64/99$.

(f) $5.\overline{97} = 5 + 97(0.\overline{01}) = 5 + 97(1/99) = 5\frac{97}{99}$.

Section 7.2

19. When mentally converting decimals into fractions, 1/999

can be substituted for the decimal $0.\overline{001}$ since they are equal.

(a) $0.\overline{003} = 3(0.\overline{001}) = 3(1/999) = 3/999 = 1/333.$

(b) $0.\overline{005} = 5(0.\overline{001}) = 5(1/999) = 5/999.$

(c) $0.\overline{007} = 7(0.\overline{001}) = 7(1/999) = 7/999.$

(d) $0.\overline{019} = 19(0.\overline{001}) = 19(1/999) = 19/999.$

(e) $0.\overline{827} = 827(0.\overline{001}) = 827(1/999) = 827/999.$

(f) $3.\overline{217} = 3 + 217(0.\overline{001}) = 3 + 217(1/999) = 3\dfrac{217}{999}.$

Section 7.2

20. (a) The number of digits in the repeating decimal is the same as the number of 9's in the denominator of the fraction.

(i) $0.\overline{23} = 23/99$ (2 digits in the repeating decimal).

(ii) $0.\overline{010} = 10/999.$

(iii) $0.\overline{769} = 769/999.$

(iv) $0.\overline{9} = 9/9 = 1.$

(v) $0.\overline{57} = 57/99 = 19/33.$

(vi) $0.\overline{1827} = 1827/9999 = 203/1111.$

(b) See the verifications below.

(i) Let $n = 0.\overline{23}$, and thus, $100n = 23.\overline{23}$.

Then $\qquad 100n = 23.\overline{23}$

$\underline{\qquad -n = \quad 0.\overline{23}}$

So $\qquad 99n = 23$

or $\qquad n = 23/99.$

(ii) Let $n = 0.\overline{010}$, and thus, $1000n = 10.\overline{010}$.

Then $\qquad 1000n = 10.\overline{010}$

$\underline{\qquad -n = \quad 0.\overline{010}}$

So $\qquad 999n = 10$

or $\qquad n = 10/999.$

(iii) Let $n = 0.\overline{769}$, and thus, $1000n = 769.\overline{769}$.

Then $\qquad 1000n = 769.\overline{769}$

$$\underline{\quad -n = \quad 0.\overline{769}\quad}$$

So $\qquad 999n = 769$

or $\qquad\quad n = 769/999.$

(iv) Let $n = 0.\overline{9}$, and thus, $10n = 9.\overline{9}$.

Then $\qquad 10n = 9.\overline{9}$

$$\underline{\quad -n = 0.\overline{9}\quad}$$

So $\qquad 9n = 9$

or $\qquad n = 1.$

(v) Let $n = 0.\overline{57}$, and thus, $100n = 57.\overline{57}$.

Then $\qquad 100n = 57.\overline{57}$

$$\underline{\quad -n = \quad 0.\overline{57}\quad}$$

So $\qquad 99n = 57$

or $\qquad n = 57/99 = 19/33.$

(vi) Let $n = 0.\overline{1827}$, and thus, $1000n = 1827.\overline{1827}$.

Then $\qquad 10000n = 1827.\overline{1827}$

$$\underline{\quad -n = \qquad 0.\overline{1827}\quad}$$

So $\qquad 9999n = 1827$

or $\qquad\quad n = 1827/9999 = 203/1111.$

Section 7.2

21. (a) Recall the pattern we discovered from earlier problems:

$0.\overline{1} = 1/9$, $0.\overline{01} = 1/99$, $0.\overline{001} = 1/999$, $0.\overline{0001} = 1/9999$. If we want five digits in the repetend, then there must be five 9's in the denominator of the

fraction. That is, $0.\overline{00001} = 1/99999$.

   (b) From earlier problems, any number of the form

$0.\overline{abcde}$, where a, b, c, d, and e are not all the same can be written as $x/99999$ where $1 \le x \le 99998$, and the digits of x are not all the same. Notice why the digits cannot all be the same. If a, b, c, d, and e all

were equal to the digit a, then $0.\overline{abcde} = 0.\overline{aaaaa} =$

$0.\overline{a}$, which has a single-digit repetend.

Section 7.2

22. $1/13 = 0.\overline{076923}$. Notice that the decimal representation has a six-digit repetend. Make a table.

Digit	0	7	6	9	2	3	0	7	6	9	2	3...
Position	1	2	3	4	5	6	7	8	9	10	11	12 ...

Every six digits, the number 3 occurs. This means that 3 is in any position that is a multiple of 6: 6, 12, 18, ...
(a) The twelfth digit is a 3, so the eleventh digit is a 2.
(b) Since $5 \times 6 = 30$, the 30th digit is 3. Therefore, the 33rd digit is a 6.
(c) Since $455 \times 6 = 2730$, the 2730th digit is a 3. Therefore, the 2731st digit is a 0.
(d) Since $1,833,333 \times 6 = 10,999,998$, the 10,999,998th digit is a 3. Therefore, the 11,000,000th digit is a 7.

Section 7.2

23. Consider $1/13$ versus $10/13$. $1/13 = 0.\overline{076923}$ and $10/13 = 0.\overline{769230}$. If we multiply the decimal expansion of $1/13$ by 10, then the decimal point is shifted one place to the right. If we ignore the first digit to the right of the decimal point in the expansion of $1/13$, then the expansions are the same. Consider $1/71 = 0.014084507...$ and $29/71 = 0.408450704...$ Since $100/71 = 1\frac{29}{71}$, we know that if we disregard the first two digits to the right of the decimal point in the decimal expansion of $1/71$, then the decimal expansions of $1/71$ and $29/71$ are the same.

Section 7.2

24. On an inexpensive, four-function calculator, the number of digits in the numerator and denominator would exceed the display capabilities of the calculator. An error symbol would be displayed if you tried to multiply out the numerator or denominator. The calculator can handle smaller calculations. Since multiplication and division can be done in any order, calculate the value by alternately dividing and multiplying .
$364 / 365 \times 363 / 365 \times 362 / 365 \times 361 / 365 \times 360 / 365 \times 359 / 365 \approx 0.943764297$. (Note: The number of decimal places displayed by your calculator may differ.)

Section 7.3

11. Grape juice is made from water mixed with juice concentrate. In the problem a "part-to-part" comparison is given: 1 part concentrate to 3 parts water. Since we want to know how much grape juice can be made from 10

ounces of concentrate, we need to set up a proportion using the ratio of concentrate to grape juice. We know that 1 part concentrate + 3 parts water = 4 parts grape juice.

$$\frac{1 \text{ part concentrate}}{4 \text{ parts juice}} = \frac{10 \text{ ounces concentrate}}{x \text{ ounces juice}}$$

Cross multiplying gives $x = 4 \times 10$

$$x = 40 \text{ ounces.}$$

Therefore, 40 ounces of grape juice can be made.

Section 7.3

12. Set up a proportion using the ratio comparing acres to days.

$$\frac{\frac{1}{2} \text{ acre}}{3 \text{ days}} = \frac{2\frac{3}{4} \text{ acres}}{x \text{ days}}$$

$$\frac{1}{2}x = 3(2\frac{3}{4})$$

$$\frac{1}{2}x = 3(\frac{11}{4})$$

$$\frac{1}{2}x = \frac{33}{4}$$

$$2x = 33$$

$$x = 16\frac{1}{2} \text{ days.}$$

It will take $16\frac{1}{2}$ days to clear the entire plot of $2\frac{3}{4}$ acres.

Section 7.3

13. Set up a proportion using the ratio comparing peaches to servings.

$$\frac{6 \text{ peaches}}{4 \text{ servings}} = \frac{x \text{ peaches}}{10 \text{ servings}}$$

$$6(10) = 4x$$

$$60 = 4x$$

$$15 = x.$$

Therefore, 15 peaches would be needed.

Section 7.3

14. Set up a proportion using the ratio comparing ounces to weeks. (Note: There are 52 weeks in a year.)

$$\frac{128 \text{ ounces}}{6\frac{1}{2} \text{ weeks}} = \frac{x \text{ ounces}}{52 \text{ weeks}}$$

$$128(52) = 6\frac{1}{2}x$$

$$6656 = 6.5x$$

$$1024 = x.$$

Therefore, 1024 ounces of liquid laundry detergent will be needed in one year.

Section 7.3

15. Since he slept 8 hours out of every 24-hour period, he slept $\frac{8}{24} = \frac{1}{3}$ of each day. Therefore, he would have spent $\frac{1}{3}$ of his life of 92 years sleeping. He slept approximately $\frac{1}{3}(92) = 30\frac{2}{3}$ years.

Section 7.3

16. Set up a proportion using the ratio comparing weight on earth to weight on the moon.

$$\frac{175 \text{ pounds (on earth)}}{28 \text{ pounds (on moon)}} = \frac{30 \text{ pounds (on earth)}}{x \text{ pounds (on moon)}}$$
$$175x = 28(30)$$
$$175x = 840$$
$$x = 4.8.$$

Therefore, the 30-pound dog would weigh 4.8 pounds on the moon.

Section 7.3

17. Set up a proportion using the ratio of miles to years.

$$\frac{4460 \text{ miles}}{0.5 \text{ years}} = \frac{x \text{ miles}}{2.75 \text{ years}}$$

Cross multiplying yields
$$4460(2.75) = 0.5(x)$$
$$12,265 = 0.5x$$
$$24,530 = x.$$

Your car will have gone 24,530 miles in 2.75 years.

Section 7.3

18. Set up a proportion using the ratio of altitude in feet to horizontal feet. Notice that the units are inconsistent in the information given. Convert miles to feet using the fact that 1 mile = 5280 feet.

$$\frac{5 \text{ feet (altitude)}}{16.37 \text{ feet (horizontal)}} = \frac{x \text{ feet (altitude)}}{5280 \text{ feet (horizontal)}}$$

Cross multiplying yields

$$5(5280) = 16.37x$$
$$26400 = 16.37x$$
$$1612.71 \approx x.$$

She gained approximately 1613 feet in altitude.

Section 7.3

19. The information about the distance the "Spruce Goose" flew is unnecessary to complete this problem. Set up a proportion using ratios comparing length of plane to wingspan. Be sure to convert all measurements to the same units.

    Plane length, 218 ft. 8 in. = 2624 inches
    Plane wingspan, 319 ft. 11 in. = 3839 inches

$$\frac{2624 \text{ inches (plane length)}}{3839 \text{ inches (plane wingspan)}} = \frac{20 \text{ inches(model length)}}{x \text{ inches (model wingspan)}}$$
$$2624x = 3839(20)$$
$$2624x = 76780$$
$$x \approx 29.26 \text{ inches.}$$

    Therefore, the wingspan of the model will be approximately 29.26 inches.

Section 7.3

20. Since the teacher : pupil ratio is 1: 35, and the school has 1400 students, set up a proportion to find out how many teachers there are currently.

$$\frac{1 \text{ teacher}}{35 \text{ pupils}} = \frac{x \text{ teachers}}{1400 \text{ pupils}}$$

    Cross multiplying yields
$$1400 = 35x$$
$$40 = x.$$
    There are currently 40 teachers.

    (a) Set up a similar proportion to find out how many teachers will be needed if we want the teacher-pupil ratio to be 1:20.

$$\frac{1 \text{ teacher}}{20 \text{ pupils}} = \frac{x \text{ teachers}}{1400 \text{ pupils}}$$

    Cross multiplying yields
$$1400 = 20x$$
$$70 = x.$$
    70 teachers will be needed, so 30 more teachers need to be hired.

    (b) If the teacher-pupil ratio is 1: 35, then from part(a), we know there are 40 teachers. The total cost for teachers is $33,000 × 40 = $1,320,000. The amount spent is $1,320,000 ÷ 1400 pupils ≈ $942.86 per pupil.

    (c) If the teacher-pupil ratio is 1: 20, then from part(b), we know there are 70 teachers. The total cost for teachers is $33,000 × 70 = $2,310,000. The amount spent is $2,310,000 ÷ 1400 pupils = $1650.00 per pupil.

Section 7.3

21. Given that ΔABC is similar to ΔDEF, we know that the ratios of corresponding sides are equal.

Correspondences		Ratio
C ↔ F	$\overline{CB}$ ↔ $\overline{FE}$	$\overline{CB}$: 15
B ↔ E	$\overline{BA}$ ↔ $\overline{ED}$	6 : $\overline{ED}$
A ↔ D	$\overline{AC}$ ↔ $\overline{DF}$	12 : 18 = 2 : 3

Since the ratios are equal, we can set up proportions to find the unknown side lengths.

Let x = length of $\overline{CB}$.   Let y = length of $\overline{ED}$.

$$\frac{2}{3} = \frac{x}{15} \qquad\qquad \frac{2}{3} = \frac{6}{y}$$
$$2(15) = 3x \qquad\qquad 2y = 3(6)$$
$$30 = 3x \qquad\qquad 2y = 18$$
$$10 = x \qquad\qquad y = 9$$

Thus, $\overline{CB}$ has a length of 10 and $\overline{ED}$ has a length of 9.

Section 7.3

22. (a) Set up a proportion using the ratio of the distance from Earth to Mars to the distance from Earth to Pluto.

$$\frac{1 \text{ (Earth to Mars)}}{12.37 \text{ (Earth to Pluto)}} = \frac{x \text{ (Earth to Mars)}}{30.67 \text{ (Earth to Pluto)}}$$

Cross multiplying yields
$$30.67 = 12.37x$$
$$2.479 \approx x.$$

Mars was 2.479 AU from Earth.

(b) Set up a proportion using the ratio of the distance in astronomical units to the distance in miles.

$$\frac{30.67 \text{ (AU Pluto to Earth)}}{2.85231 \times 10^9 \text{ (miles Pluto to Earth)}} = \frac{1 \text{ (AU Earth to Sun)}}{x \text{ (miles Earth to Sun)}}$$

Cross multiplying yields
$$30.67x = 2.85231 \times 10^9$$
$$x = 93,000,000.$$

The Earth is about 93,000,000 miles from the Sun.

(c) From part(a), we know that Mars was 2.479 AU from Earth in October 1985. Since we know how miles and AU are related from part(b), we can set up a proportion using the ratio of the distance in miles to the distance in astronomical units.

$$\frac{93000000 \text{ (miles)}}{1 \text{ (AU)}} = \frac{x \text{ (miles)}}{2.479 \text{ (AU)}}$$

Cross multiplying yields:
$$93,000,000(2.479) = x$$
$$230,547,000 = x.$$

In October 1985, Mars was about 2.31×10^8 miles from Earth.

Section 7.3

23. Set up a proportion in each part.
    (a) We are comparing years to hours in our analogy.

    $$\frac{10^{10} \text{ years}}{24 \text{ hours}} = \frac{x \text{ years}}{1 \text{ hour}}$$

    Cross multiplying yields
    $$10^{10} = 24x$$
    $$x \approx 416,666,667.$$

    One hour corresponds to approximately 416,666,667 years of actual time.

    (b) Since from part(a) we know how one hour is related to actual years, we can set up a proportion using 60 minutes rather than 1 hour.

    $$\frac{416666667 \text{ (years)}}{60 \text{ (minutes)}} = \frac{x \text{ (years)}}{1 \text{ (minute)}}$$

    Cross multiplying yields
    $$416666667 = 60x$$
    $$x \approx 6,944,444.$$

    One minute corresponds to approximately 6,944,444 years of actual time.

    (c) Since from part(b) we know how one minute is related to actual years, we can set up a proportion using 60 seconds rather than 1 minute.

    $$\frac{6944444 \text{ years}}{60 \text{ seconds}} = \frac{x \text{ years}}{1 \text{ second}}$$

    Cross multiplying yields
    $$6944444 = 60x$$
    $$x \approx 115,741.$$

    One second corresponds to approximately 115,741 years of actual time.

(d) If Earth was formed approximately 5 billion years ago, then find what that length of time this corresponds to in hours. From part (a), we know that 1 hour corresponds to 416,666,667 years.

$$\frac{1 \text{ hour}}{416666667 \text{ years}} = \frac{x \text{ hours}}{5000000000 \text{ years}}$$

Cross multiplying yields
$$5{,}000{,}000{,}000 = 416{,}666{,}667x$$
$$x = 12.$$
Since the 24-hour day begins at midnight, the earth was formed at about 12 noon.

(e) From part (b), we know that 1 minute corresponds to 6,944,444 years.

$$\frac{1 \text{ minute}}{6944444 \text{ years}} = \frac{x \text{ minutes}}{2600000 \text{ years}}$$

Cross multiplying yields
$$2600000 = 6944444x$$
$$x = 0.3744.$$

Notice $(0.3744 \text{ minutes})\left(\dfrac{60 \text{ seconds}}{1 \text{ minute}}\right) \approx 22.5$ seconds.

Since the 24-hour day begins at midnight, the creatures who left these remains died close to 24 hours later, at about 22.5 seconds before midnight.

(f) From part (c), we know that 1 second corresponds to 115,741 years.

$$\frac{1 \text{ second}}{115741 \text{ years}} = \frac{x \text{ seconds}}{10000 \text{ years}}$$

Cross multiplying yields
$$10{,}000 = 115{,}741x$$
$$x = 0.0864.$$

Thus, the growth of modern civilization began 0.0864 seconds before midnight.

Section 7.3

24. Let d = Distance to the airport. Recall the relationship Distance = Rate × Time. Since we want to find the distance and we know the rate, we solve the equation for time so that it can be eliminated: Time = Distance/Rate. If Cary drives at different rates, then his travel times will differ. The time he takes when driving 60 mph, say t_1, is $t_1 = d/60$. The time he takes when driving 30 mph, say t_2, is $t_2 = d/30$. Notice that there is a two-hour difference in arrival times. Since traveling 30 mph takes 2 hours longer

we have $t_2 = t_1 + 2$. Substituting, we get $d/30 = d/60 + 2$. Multiplying each term by 60, the least common denominator, yields $2d = d + 120$, so $d = 120$. Thus, the airport was 120 miles away.

Section 7.3

25. Working backwards, we know the man had no money after he bought the shoes. In order to buy the shoes, he had to have exactly $20, half of which was given to him by his father. Therefore, the man had $10 in his pocket after he bought the slacks. Now in order to buy the slacks *and* have $10 left over, the man must have had $30, half of which was given to him by his father. Therefore, the man had $15 in his pocket after he bought the hat. Finally, in order to buy the hat *and* have $15 left over, the man must have had $35, half of which was given to him by his father. Therefore, the man had $17.50 when he walked into the store.

    Another way to solve this problem is to use variables. Let $x =$ the amount of money the man had when he walked into the store. His father gave him x dollars, and the man bought a $20 hat. The man had $2x - 20$ dollars after buying the hat. His father gave him $2x - 20$ dollars, and the man bought $20 slacks. The man had $2x - 20 + 2x - 20 - 20 = 4x - 60$ dollars left after buying the slacks. His father gave him $4x - 60$ dollars, and the man bought $20 shoes. The man had no money left at this point.

    $$4x - 60 + 4x - 60 - 20 = 0$$
    $$8x - 140 = 0$$
    $$8x = 140$$
    $$x = 17.50.$$

    Therefore, the man had $17.50 when he walked into the store.

Section 7.3

26. Notice that in order to answer this question, it is necessary to exclude silver dollars or the answer will be infinitely large just by using all silver dollars. If we do not have change for a nickel, we could have at most 4 pennies. If we do not have change for a dime, we could have at most 1 nickel and 4 pennies. If we do not have change for a quarter, we could have at most 1 dime, 1 nickel, and 4 pennies, or 2 dimes and 4 pennies. (Recall that we want the largest sum of money when we are done, so we choose the 2 dimes and 4 pennies option.) If we do not have change for a half-dollar, we could have at most 1 quarter, 4 dimes, and 4 pennies. If we do not have change for a dollar, we could have at most 1 half-dollar, 1 quarter, 4 dimes, and 4 pennies. With this combination, we have

119 cents without being able to give change for a nickel, a dime, a quarter, a half-dollar, or a dollar.

Section 7.3

27. Notice that a single-digit number, 1 - 9, is subtracted during a turn. Consider the winning strategy for a game beginning with 20. If the loser is the one who ends at zero, you want to leave your opponent with 1 after your last turn so that she/he will be forced to subtract 1 and lose. Consider further that if, at the end of a turn, you leave your opponent with 11, no matter what digit she/he subtracts, you can always subtract enough to leave your opponent with 1, thus forcing her/him to lose. Therefore, for the game beginning at 20, you can always win by going first and subtracting 9. Your opponent will be left with 11. No matter what your opponent subtracts, you can subtract enough to leave her/him with 1, which she/he must subtract, and thus lose.

In the game beginning with 100, the first player should always subtract 9, leaving 91. On successive turns, the first player should subtract whatever is necessary to leave 81, 71, 61, 51, 41, 31, 21, 11, and 1, thus forcing the opponent to subtract 1 and lose.

Section 7.3

28. Notice there is a 4-minute difference between the two timers. In order to cook something for exactly 15 minutes, we need to use the 11-minute timer and this 4-minute difference between the timers. Begin both timers at the same time. When the 7-minute timer runs out, begin cooking since there are 4 minutes left on the 11-minute timer. After we have been cooking for 4 minutes, the 11-minute timer will run out. Turn it over and continue cooking until it runs out. If we stop cooking at this point, we will have been cooking for 15 minutes.

Section 7.3

29.

      1   2   3   4   5   6   7   8   9 10 11 12
                          Posts

The distance from Post 1 to Post 2 is one "post length". Since the runner began at the first post, she traveled only seven post lengths by the time she reached the eighth post. Since we want to know how many seconds it will take to reach the twelfth post, we need to set up a proportion. Notice that by the time she reaches the twelfth post she will have traveled eleven post lengths.

$$\frac{8 \text{ seconds}}{7 \text{ post lengths}} = \frac{x \text{ seconds}}{11 \text{ post lengths}}$$
$$88 = 7x$$
$$12\frac{4}{7} = x.$$

It will take $12\frac{4}{7}$ seconds to reach the twelfth post.

Section 7.3

30. Notice that each child had a different number of pennies. Since the ratio of each child's total to the next richer child's total is a whole number, every child's total must be divisible by each of the poorer children's totals. If all the amounts were added, the poorest child's total could be factored out of the sum. All together they had 2879 cents. Since the poorest child's total can be factored out, we know that (poorest child's total) $\times$? = 2879. Notice that 2879 is prime, so 2879 = 1×2879. Therefore, the poorest child must have only 1 cent.

The remaining six children must have 2878 cents all together. Once again, each amount is divisible by the second poorest child's total. Since 2878 = 2×1439 and 1439 is prime, the second poorest child must have 2 cents. Continue in this manner. The remaining 5 children have 2876 cents. Since 2876 = 4×719, the third poorest child must have 4 cents. The remaining 4 children have a total of 2872 cents. Since 2872 = 8×359, the fourth poorest child must have 8 cents. The 3 remaining children have a total of 2864 cents. Since 2864 = 16×179, the third richest child has 16 cents. The remaining 2 children have 2848 cents. Since 2848 = 32×89, the second richest child has 32 cents. There are only 2816 cents left, which belong to the one remaining child. Therefore, the totals for each of the seven children are 1, 2, 4, 8, 16, 32, and 2816 cents.

Section 7.3

31. To be sure of getting at least 2 of one kind, consider what could happen in the worst case. Since there are 3 kinds of apples, we could get one of each in three draws. Finally, on the fourth draw, we would be sure to have 2 of one kind. To be sure of getting at least 3 of one kind, we must have 7 apples. This is because before we draw a third apple of one kind, we might have 2 of each of the three kinds, and 2(3 kinds) + 1 = 7. To get at least 10 of one kind, we need to take 28 apples since before we draw a tenth apple of one kind, we could have 9 of each of the three kinds and 9(3 kinds) + 1 = 28.

Now we make a list and look for a pattern.

# of One Kind	Draws
2	$4 = 3 \times 1 + 1$
3	$7 = 3 \times 2 + 1$
10	$28 = 3 \times 9 + 1$
n	$3n - 2 = 3(n - 1) + 1$

To draw at least n of one kind, we must take $3n - 2$ apples.

Section 7.4
18. (a) Recall the formula for simple interest:
$$\text{Interest} = \text{Principal} \times \text{Rate}.$$
Let $p =$ the amount of principal. In this case, we know the interest earned and the rate. (Note: 4.25% = 0.0425.) Therefore, $p\,(0.0425) = 208.76$, and $p = 4912$. The total amount in the account at the end of the year is the principal + interest = $4912 + 208.76 = \$5,120.76$.
   (b) At the rate of 5.33% = 0.0533, the investor would have earned $4912(0.0533) = \$261.81$ in interest. The investor would have earned $\$261.81 - \$208.76 = \$53.05$ more at the rate of 5.33%.

Section 7.4
19. Suppose you "borrowed" $100 by purchasing a vacuum cleaner on credit. You have two options: you can pay the $100 now, or you can wait 15 days and pay a daily interest rate of 0.04839%. How much will you save by paying the "loan" 15 days before it is due?

Pay now: Amount due = $100.
Pay in 15 days: Amount due = $p(1+ r)^t$, where $p =$ principal, $r =$ daily interest rate, and $t =$ time in days. Interest is compounded daily for 15 days. Therefore,

$$\begin{aligned} \text{Amount due} &= 100(1+ 0.0004839)^{15} \\ &\approx 100\,(1.0073) \\ &\approx 100.73. \end{aligned}$$

You will save about 73 cents by paying now rather than in 15 days.

Section 7.4
20. They lost 2 out of 35 games, so the fraction of games lost is $\frac{2}{35}$. Since $\frac{2}{35} = 0.057$, the percent of games lost is 5.7%.
Therefore, the percent of games won is $100\% - 5.7\% = 94.3\%$.

Section 7.4

21. (a) In 1991 the increase in contributions was 6.2% of the total 1990 contribution. Let c = the contribution for 1990. Then the total for 1991 was 106.2% c , or 1.062c.

$$1.062c = 1.25 \times 10^{11}$$
$$c \approx 1.18 \times 10^{11}.$$

Therefore, the amount donated to charity in 1990 was about 1.18×10^{11}.

(b) The percent of all charitable contributions by individuals can be found by multiplying the ratio $\dfrac{\text{Contributions by individuals}}{\text{Total contributions}}$ by 100%.

$$\frac{1.03 \times 10^{11}}{1.25 \times 10^{11}} \times 100\% = (0.824)(100\%) = 82.4\%.$$

Thus, 82.4% of all charitable contributions were made by individuals.

Section 7.4

22. (a) Notice that each source of energy in the pie graph is given in quadrillion BTU's. Add each amount to find the total amount of energy produced.

$$6.1 + 20.9 + 5.7 + 17.3 + 19.5 = 69.5.$$

Therefore, 69.5 quadrillion BTU's of energy were produced, from all sources, in the United States, in 1988.

(b) To find the percent of the energy produced from each source, divide each source by the total amount of energy produced and multiply by 100%.

Renewables: $\dfrac{6.1}{69.5} \times 100\% \approx 8.8\%$

Coal: $\dfrac{20.9}{69.5} \times 100\% \approx 30.1\%$

Nuclear: $\dfrac{5.7}{69.5} \times 100\% \approx 8.2\%$

Crude Oil: $\dfrac{17.3}{69.5} \times 100\% \approx 24.9\%$

Natural Gas: $\dfrac{19.5}{69.5} \times 100\% \approx 28.1\%$

Notice that the percentages do not add to 100% due to rounding.

Section 7.4
23. The price was not consistent with the ad. If the original price was discounted by 15%, then the price should be lowered by ($115)(0.15) = $17.25. Therefore, the new price should be $115.00 – $17.25 = $97.75. The sale price that was offered was $100, which was $15 off the original price. What percent of $115 is this $15? We can now write an equation as shown next.

$$15 = x(115)$$
$$15 \div 115 = x$$
$$0.1304 \approx x.$$

The advertised discount was really 13.04%.

Section 7.4
24. Since 17% of the selling price is $850, set up an equation to find the selling price. Let x = the selling price of the car. Then

$$0.17x = 850$$
$$x = 850 \div 0.17 = 5000.$$

Therefore, the car sold for $5000.

Section 7.4
25. Notice a relationship between the following two statements: 20% of 50 is 10, and 10% of 100 is 10. If we divide the percent by 2, then we multiply the other number by 2. This can be done mentally.
   (a) Since the percent is divided by 5, multiply 50 by 5. 30% of 50 is 6% of 250.
   (b) Since the percent is divided by 8, multiply 60 by 8. 40% of 60 is 5% of 480.
   (c) Since the number has been doubled, the percent should be halved. 30% of 80 is 15% of 160.

Section 7.4
26. Let x = the original price of the car. Since there is an 8% discount, we paid 92% of the original price of the car. We paid $4485.00, so $4485.00 = 92% × x. Thus, 4485 = 0.92x, so 4875 = x. The original price of the car was $4875.00.

Section 7.4
27. (a) To find the percent of the surface area of the earth these continents comprise, consider the ratio that compares the surface areas of these continents to the surface area of the earth. Multiply that ratio by 100%.

$$\frac{8.5 \times 10^{13}}{5.2 \times 10^{14}} \times 100\% \approx 16.3\%.$$

The continents comprise about 16.3% of the earth's surface area.

(b) Consider the ratio comparing the surface area of the Pacific Ocean to the surface area of the earth.

$$\frac{1.81 \times 10^{14}}{5.2 \times 10^{14}} \times 100\% \approx 34.8\%.$$

The Pacific Ocean covers about 34.8% of the surface of the earth.

(c) If oceans cover 70% of the surface area of the earth, multiply to find how many square meters of the earth's surface are covered by oceans.

$$(0.70)(5.2 \times 10^{14}) = 3.64 \times 10^{14} \text{ m}^2.$$

(d) Since from part (c), the oceans make up 70% of the surface area, we can state that land makes up the other 30% of the surface area. The total surface area of the earth is 5.2×10^{14} m^2, so 30% of 5.2×10^{14} = $(0.30)(5.2 \times 10^{14}) = 1.56 \times 10^{14}$ m^2. The percent contained in the state of Texas is

$$\frac{6.92 \times 10^{11}}{1.56 \times 10^{14}} \times 100\% \approx 0.44\%,$$

or approximately $\frac{11}{25}$ of 1%.

Section 7.4

28. The information provided is not consistent. Let x = the number of grams of protein recommended. If 3 grams are 4% of the U.S. RDA, then set up a proportion to find out how many grams correspond to 100% U.S.RDA.

$$\frac{3 \text{ grams}}{4\%} = \frac{x \text{ grams}}{100\%}$$
$$300 = 4x$$
$$75 = x.$$

Therefore, 75 grams of protein are recommended daily. On the other hand, if 7 grams are 15% of the U.S. RDA, then 100% of the U.S. RDA corresponds to $46\frac{2}{3}$ grams of protein, as shown below.

$$\frac{7 \text{ grams}}{15\% \text{ RDA}} = \frac{x \text{ grams}}{100\% \text{ RDA}}$$

$$700 = 15x$$
$$46\frac{2}{3} = x.$$

The U.S. RDA of protein cannot be both 75 grams and $46\frac{2}{3}$ grams.

Section 7.4

29. The original inseam is 32 inches. If the inseam is made 10% longer, then it is increased by 32(0.10) = 3.2 inches. The new inseam is then 32 + 3.2 = 35.2 inches. The material is expected to shrink by 10%. So, the slacks will shrink by 35.2(0.1) = 3.52 inches. The final inseam measurement will thus be 35.2 – 3.52 = 31.68 inches.

Section 7.4

30. Suppose that an item costs $100. If the price is marked up by 10%, then the price is increased by 100(0.10) = $10 for a new price of $100 + $10 = $110. Then, if the item is discounted by 10%, the price is decreased by 110(0.10) = $11 for a final price of $110 – $11 = $99.

    On the other hand, if the price of the item is first decreased by 10%, then the price is decreased by 100(0.10) = $10 for a new price of $100 – $10 = $90. Then, if the price is marked up by 10%, then the price is increased by 90(0.10) = $9 for a final price of $90 + $9 = $99. We see that the final price of $99 is the same either way.

    Another way to look at this problem is to use a variable. Let x = the original price of the item. Method 1 calls for a 10% markup followed by a 10% discount.

    Price after 10% markup: $x + 0.10x = 1.10x$
    Price after 10% discount: $1.10x - 1.10x(0.10) = 0.99x$

    Method 2 calls for a 10% discount followed by a 10% markup.

    Price after 10% discount: $x - 0.10x = 0.90x$
    Price after 10% markup: $0.90x + 0.90x(0.10) = 0.99x$

    By either method, the final price is 99% of the original price.

Section 7.4

31. Suppose Cathy has 100 baseball cards. Since percent means "per hundred", and Joseph has 64% as many cards as Cathy, Joseph has 64 cards. Martin has 50% , or half as many cards as Joseph. Therefore, Martin has 64 ÷ 2 = 32 cards. Martin has 32 cards compared to Cathy's 100 cards. Since 32 ÷ 100 = 32%, Martin has 32% as many cards as Cathy.

Section 7.4

32. For a 50-year-old man, subtract 50 from 220: 220 – 50 = 170. Find 70% of this difference: (0.70)(170) = 119. Now find 80% of the difference: (0.80)(170) = 136. The optimal

heart rate for a 50-year-old man is between 119 and 136 beats per minute.

Section 7.4
33. Let x = the original number of outputs of audio information provided by the competition. If 6 outputs is 40% more than the competition, then 140% of x should be 6. Thus, we have

$$140\% \text{ of } x \text{ is } 6$$
$$1.4x = 6$$
$$x \approx 4.2857.$$

We have found that the number of outputs of audio information provided by the competition is about 4.2857. However, since outputs of audio information must be whole numbers, the number of outputs *cannot* be 4.2857. Therefore, 6 is not 40% more than the competition.

Section 7.4
34. Compare these billing rates by supposing the doctor charges you $100. By the first method, 15% would be taken off the bill. 15% of 100 = 0.15(100) = 15, so the bill would be $100 − $15 = $85. By the second method, 10% is taken off first and then 5% is taken off the discounted amount. 10% of 100 = (0.10)(100) = 10, for a bill of $100 − $10 = $90. 5% of 90 = 0.05(90) = 4.5, so the final bill would be $90 − $4.5 = $85.50. You would save money by choosing to have the discounts added, as in the first method.

Section 7.4
35. Let x = the population of the country in 1990. Since the population increased by 4.2% during 1990, there were 104.2%(x), or 1.042x, people at the beginning of 1991. Since the population increased by 2.8% during 1991, there were 102.8%(1.042x) = (1.028) (1.042x) = 1.071176x people at the beginning of 1992. Since the population *decreased* by 2.1% during 1992, there were 97.9%(1.071176x) = (0.979)(1.071176x) ≈ 1.0486813x people at the beginning of 1993. Therefore, there was an increase of approximately 4.9% over the 3 year period.

Section 7.4
36. You cannot take more than two petals so when it is your turn, you do not want to be left with three petals. If, on your turn, you are faced with three petals, no matter how many you take (one or two), your opponent can take the final one or two petals to win the game. Similarly, you never want to be left with six petals because no matter what you take (one or two petals), your opponent can take one or two petals to leave you with three petals - a losing

situation for you. Finally, if you go first when there are nine petals, then you could be forced to lose if your opponent leaves you with six petals and then with three petals on your next turn. The person who begins the game can always be forced to lose. To win, a player should start second and always leave the other player with a multiple of three petals. However, if the player who goes second does not finish a turn by leaving a multiple of three petals, the first player can end up a winner.

Section 7.4
37. Suppose an item was originally priced at $100. By method (i), 20% is deducted first and then a 6% tax is added. Notice that a 20% markdown is the same as paying 80% of the original price. 80% of 100 = 0.80(100) = $80. Adding a 6% tax gives 106% of 80 = 1.06(80) = $84.80. By method (ii), the 6% tax is added first and then 20% is deducted. 106% of 100 = 1.06(100) = $106. Paying 80% of this gives 80% of 106 = 0.80(106) = $84.80. Therefore, both methods are the same. To see why this is true, let p = the original price of the item. The final price by method (i) is found by $p \times 80\% \times 106\%$. The final price by method (ii) is found by $p \times 106\% \times 80\%$. Since multiplication is commutative, both methods yield the same result.

Section 7.4
38. Use the compound interest formula $P(1 + r/n)^{nt}$. Since semi-annual compounding means interest is compounded twice a year, $n = 2$. The principal borrowed is $12,000, so $P = 12000$. The annual interest rate is 7%. Therefore, $r = 0.07$. The time is 3 years, so $t = 3$.

$$P(1 + r/n)^{nt} = 12000(1 + 0.07/2)^{(2)(3)}$$
$$= 12000(1.035)^6$$
$$\approx 12000(1.229)$$
$$\approx 14751.06.$$

Her tax-deferred account will be worth approximately $14,751.06 at the end of 3 years.

Section 7.4
39. Each year she needs to earn 11% more than she did during the previous year. Therefore, one year from now, she needs to earn 111% of her current salary or 1.11(35,000). Two years from now, she needs to earn 111% of her previous year's salary or 1.11(1.11)(35,000) = $(1.11)^2(35,000)$. Three years from now, she needs to earn 111% of her previous year's salary; that is, she must earn $(1.11)(1.11)^2(35000) = (1.11)^3(35,000)$. Four years from now, she needs to earn $(1.11)^4(35,000)$. Five years from

now, she needs to earn $(1.11)^5(35,000)$. Therefore, in order for her to have the same buying power in five years, she must earn $(1.11)^5(35,000) \approx 58977.04$, or approximately $59,000.

Section 7.4

40. Let x = the initial investment. At the end of year 1, they will have 8.25% more than the initial investment or 108.25% of the initial investment. ($108.25\% \ x = 1.0825x$.) Each year the amount is 108.25% of the previous year's balance or 1.0825 times as much.
    After 10 years, this will have been done 10 times, so the amount in the account will be $(1.0825)^{10}x$. Since we want the account to contain $20,000, we must have
    $$(1.0825)^{10}x = 20,000$$
    $$2.209x \approx 20,000$$
    $$x \approx 9052.13.$$
    Therefore, the initial investment must be $9052.13.

Section 7.4

41. The 1982-1983 year was the base year for comparison, so the CPI for it was 100%. In calculating the percent increase from 1985 to 1986, we consider the change from 1985 to 1986 as compared to what it was in 1985.
    $$\% \text{ Increase} = \frac{\text{CPI for 1986} - \text{CPI for 1985}}{\text{CPI for 1985}} \times 100\%$$
    $$= [(115.7 - 111.2) \div 111.2] \times 100\%$$
    $$\approx (4.5 \div 111.2) \times 100\%$$
    $$\approx 4.05\%.$$
    The increase in the CPI from 1985 to 1986 was approximately 4.05%.

Section 7.4

42. If your income is low, then the percent taken in taxes is low. If your income is high, then the percent taken in taxes is high. Make a table of salaries, taxes, and net earnings, and find the salary that lets you keep more of your money.

Salaries (dollars)	Tax (percent)	Tax (dollars)	Net Salary (dollars)
10,000	10	1,000	9,000
20,000	20	4,000	16,000
30,000	30	9,000	21,000
40,000	40	16,000	24,000
50,000	50	25,000	25,000
60,000	60	36,000	24,000
70,000	70	49,000	21,000
80,000	80	64,000	16,000
90,000	90	81,000	9,000

Notice the symmetry in the net salaries. At $50,000, your net salary is a maximum at $25,000.

SOLUTIONS - PART A PROBLEMS

Chapter 8: Integers

Section 8.1

19. (a) The set of integers is closed under subtraction. For the set to be closed under subtraction, we must be able to take any two integers, find their difference, and obtain an integer. As an example, notice that 6 and 8 are integers. Also, $6 - 8 = -2$ and -2 is an integer.

   (b) Integer subtraction is not commutative. We cannot subtract integers in any order and obtain the same result. For example, $2 - 6 = -4$, but $6 - 2 = 4$.

   (c) Integer subtraction is not associative. As a counterexample, consider the following:
   $$(10 - 4) - 7 \neq 10 - (4 - 7)$$
   $$6 - 7 \neq 10 - (-3)$$
   $$-1 \neq 13.$$

   (d) There is no identity element for integer subtraction. As a counterexample, consider the following:
   $$0 - 6 \neq 6 - 0$$
   $$-6 \neq 6.$$

   Assume that n is the identity element for subtraction. Then it must be true that $a - n = n - a = a$, for any integer a. If $a - n = n - a$, then $a - n = -(a - n)$, but this can only be true when $a - n = 0$. Since we have that $a - n = a$, and $a - n = 0$, it follows that $a = 0$. However, $a - n = n - a = a$ must be true for any a and not just for $a = 0$. Therefore, there is no identity element for integer subtraction.

Section 8.1

20. Suppose the adding-the-opposite approach is true. That is, suppose that $a - b = a + (-b)$. We must show that if $a - b = c$, then $a = b + c$. Suppose that $a - b = c$.

Then	$a + (-b) = c$	Adding-the-opposite approach
	$a + (-b) + b = c + b$	Add b to both sides
	$a = b + c$.	Additive inverse, commutative, and identity properties

   Therefore, if $a - b = c$, then $a = b + c$.
   Now we must show that if $a = b + c$, then $a - b = c$.
   Suppose that $a = b + c$.

Then	$a + (-b) = b + c + (-b)$	Add $(-b)$ to both sides.
	$a + (-b) = c$	Commutative, additive inverse, and identity properties.
	$a - b = c$.	Adding-the-opposite approach.

So, if $a = b + c$, then $a - b = c$.

Therefore, we see that the missing-addend approach is a consequence of the adding-the-opposite approach.

Section 8.1

21. (a) (i) Consider the following examples:

If $a = 3$ and $b = 11$, then the equation is true.

$|3 + 11| = |14| = 14$, and

$|3| + |11| = 3 + 11 = 14$.

Therefore, $|3 + 11| = |3| + |11|$.

If $a = -13$ and $b = -2$, then the equation is true.

$|-13 + (-2)| = |-15| = 15$, and

$|-13| + |-2| = 13 + 2 = 15$.

So $|-13 + (-2)| = |-13| + |-2|$.

If $a = -6$ and $b = 4$, then the equation is false.

$|-6 + 4| = |-2| = 2$, and

$|-6| + |4| = 6 + 4 = 10$.

But, since $2 \neq 10$, $|-6 + 4| \neq |-6| + |4|$.

In general, if a and b have the same sign, or if one or both of them are zero, then it is true that $|a + b| = |a| + |b|$.

(ii) Recall the third example from part (i). The numbers have opposite signs. $|-6 + 4| = 2 < 10 = |-6| + |4|$. Notice that when a and b are nonzero and have opposite signs, the quantity $|a + b|$ is the absolute value of a difference, while the quantity $|a| + |b|$ is the sum of two positive numbers. Since adding integers with different signs is like subtracting, the quantity $|a + b|$ will always be less than the quantity $|a| + |b|$. In general, if a and b are both nonzero and have opposite signs, then $|a + b| < |a| + |b|$.

(iii) The inequality $|a + b| > |a| + |b|$ is never true. Consider the various possibilities for integers a and b. They could have the same signs (positive or negative), different signs, both zero, or one zero. Recall what we did in parts (i) and (ii). If a and b have the same sign or one or both of them are zero, then $|a + b| = |a| + |b|$. If a and b have different signs, then $|a + b| < |a| + |b|$. Since there are no other possibilities for a and b, there is no case where $|a + b| > |a| + |b|$.

(iv) Recall the discussion from part (iii). For any combination of integers a and b, either $|a + b| = |a| + |b|$ or $|a + b| < |a| + |b|$. Therefore, we have $|a + b| \leq |a| + |b|$, and the inequality holds for all integers.

(b) Recall our discussion from part (a) above. The inequality $|a + b| \leq |a| + |b|$ will hold for all pairs of integers.

Section 8.1

22. (a) The set A is closed under subtraction. We know that 4 and 9 are elements of A. Recall that we can take any two elements of A, not necessarily different, subtract them and have an element in the set.

(i)   Since $9 - 4 = 5$, 5 is an element of A.

(ii) Since $4 - 9 = -5$, -5 is an element of A.

(iii) Since $4 - 4 = 0$, 0 is an element of A.

(iv) Since $9 - [(4 - 4) - 4] = 13$, 13 is an element of A.

(v)   Since $(9 - 4) - 4 = 1$, 1 is an element of A.

(vi) Since $((9 - 4) - 4) - 4 = -3$, -3 is an element of A.

(b) Notice from part (a) that 1 and 0 are elements of set A. Using 1 and 0, generate more elements in the set. Note $0 - 1 = -1$, so -1 is in the set. Also, $-1 - 1 = -2$, so -2 is in the set, and $-2 - 1 = -3$, so -3 is in the set. Continuing in this manner will generate all of the negative integers. We can use the negative integers to generate the positive integers. For example, $1 - (-1) = 2$, $2 - (-1) = 3$, $3 - (-1) = 4$, and so on. Since the set contains all of the positive and negative integers, and zero, set A contains all of the integers.

(c) Since 8 is a multiple of 4, subtracting 4 repeatedly from 8 generates the elements 4, 0, -4, -8, -12, and so on. Notice that these numbers are all multiples of four. If we subtract -4 repeatedly from 8, we generate the elements 12, 16, 20, and so on. These numbers are also multiples of 4. Therefore, set A will contain all multiples of 4 (and only multiples of 4).

(d) Notice that in part(a), GCF(4,9) = 1. In part(b), GCF(4,8) = 4. In general, if a set contains two numbers whose GCF is different from 1, then the set contains only multiples of that GCF. Try more examples to convince yourself that this is true.

Section 8.1

23. In row three, we can fill in the first two boxes. The first box contains 4 since $-9 + 13 = 4$. The second box contains -2 since $13 + (-15) = -2$. The sum of the numbers in these two boxes is 2. Therefore, the first box in row two contains a 2. Now since the top box contains -22, and one of the two boxes beneath it contains 2, the other box beneath it must contain the solution to the equation $2 + \square = -22$. Therefore, the second box in row two contains -24. Similarly, the third box in row three must contain the solution to $-2 + \square = -24$, which is -22. The fourth box in row four must contain the solution to the equation $-15 + \square = -22$, which is -7. The solution is shown next.

```
            ┌─────┐
            │ -22 │
        ┌───┼─────┤
        │ 2 │ -24 │
    ┌───┼───┼─────┤
    │ 4 │-2 │ -22 │
┌───┼───┼───┼─────┤
│-9 │13 │-15│ -7  │
└───┴───┴───┴─────┘
```

Section 8.1

24. Congratulate the student on the technique. It will always work. Consider a general subtraction problem involving two-digit numbers $ab - cd$. In expanded form, this problem can be written as $(10a + b) - (10c + d)$. In the student's algorithm, he or she subtracts d from b to obtain $b - d$ and then subtracts $10c$ from $10a$ to obtain $10a - 10c$. These two results are then added, which yields the result $10a - 10c + b - d$. This result is the same as $(10a + b) - (10c + d) = 10a - 10c + b - d$. For the given example, we have $72 - 38 = (70 + 2) - (30 + 8) = (70 - 30) + (2 - 8) = 40 + (-6) = 34$.

Section 8.2

23. Consider the first square in the bottom row. Since the number in a square is the product of the two numbers beneath it, we need to solve the equation __ × 4 = –20 to find the missing number. The missing number must be –5. Consider the third square of the bottom row. It must be the solution to __ × –5 = –15. Thus, the missing number must be 3. Now we can determine the number in the middle box in the third row. Since it is the product of the numbers beneath it, we know that it is 4 × 3 or 12. At this point, the pyramid of squares looks like:

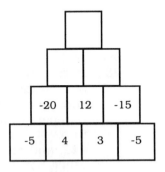

Consider the two empty squares in row two. The first must contain –20 × 12 = –240, and the second one must contain 12 × –15 = –180. Finally the top number is –240 × –180 = 43,200. Therefore, the final pyramid of squares is as follows:

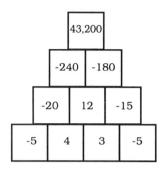

Section 8.2

24. (a) Substitute each of the integers for *x* in the inequality.

$$3(-6) + 5 = -13$$
−13 is not< −16
False when *x* = −6

$$3(-10) + 5 = -25$$
$$-25 < -16$$
True when *x* = −10

$$3(-8) + 5 = -19$$
$$-19 < -16$$
True when *x* = −8

$$3(-7) + 5 = -16$$
−16 is not< −16
False when *x* = −7

(b) Solve the inequality for *x*.

$$3x + 5 < -16$$
$$3x < -21$$
$$x < -7.$$

The inequality is true for all values of *x* less than 7. Therefore, the inequality is true when *x* is −8, −9, −10, −11, ... The largest of these integers is −8.

(c) There is not a smallest integer value of *x* that makes the inequality true. Notice that −8 is the largest integer that makes the inequality true. The inequality is also true for *x* = −9, *x* = −10, *x* = −11, *x* = −12, and so on. Continue substituting smaller and smaller integers for *x*. The inequality is true for all integers such that *x* ≤ −8.

Section 8.2

25. (a) The addition table is shown below.

+	+	−
+	+	?
−	?	−

Positive + Positive = Positive (+ sign in the table).
Negative + Negative = Negative (− sign in the table).
Negative + Positive = Positive or Negative or Zero (thus the ? sign in the table).

(b) (i) The table for subtraction is shown below.

−	+	−
+	?	+
−	−	?

Positive − Positive = Positive or Negative or Zero (thus the ? sign in the table).

Positive − Negative = Positive (+ sign in the table) [since subtracting a negative is the same as adding a positive].

Negative − Positive = Negative (− sign in the table).

Negative − Negative = Positive or Negative or Zero (thus the ? sign in the table).

(ii) The table for multiplication is shown below.

×	+	−
+	+	−
−	−	+

Positive × Positive = Positive (+ sign in the table).

Positive × Negative = Negative. (− sign in the table). [Since a and b are positive, $-b$ is negative. $(a)(-b) = -(ab)$. ab is positive, so $-ab$ is negative.]

Negative × Positive = Negative (− sign in the table).

Negative × Negative = Positive (+ sign in the table). [Since a and b are positive, $-a$ and $-b$ are negative. $(-a)(-b) = ab$. ab is positive.]

(iii) Division

÷	+	−
+	+	−
−	−	+

Positive ÷ Positive = Positive (+ sign in the table).

Positive ÷ Negative = Negative (− sign in the table).

Negative ÷ Positive = Negative (− sign in the table).

Negative ÷ Negative = Positive (+ sign in the table).

Section 8.2

26. We must calculate the number of grams in one atom of carbon. That is, we want the number of grams *per* atom. To determine this number, we divide grams by atoms.

$$\frac{12.01 \text{ grams}}{6.022 \times 10^{23} \text{ atoms}} \approx 1.99 \times 10^{-23} \text{ grams/atom.}$$

Section 8.2

27. (a) Express the rate of hair growth in scientific notation, and then convert the rate from seconds to days. The rate of hair growth is 4.3×10^{-9} meters per second.

$$\frac{4.3 \times 10^{-9} \text{ meters}}{1 \text{ second}} \times \frac{60 \text{ seconds}}{1 \text{ minute}} \times \frac{60 \text{ minutes}}{1 \text{ hour}} \times \frac{24 \text{ hours}}{1 \text{ day}} = 3.7152 \times 10^{-4} \frac{\text{meters}}{\text{day}}.$$

In one month of 30 days, the amount of hair growth would be 30 times as much as for one day.

$$\frac{3.7152 \times 10^{-4} \text{ m}}{\text{day}} \times \frac{30 \text{ days}}{\text{month}} \approx \frac{1.11 \times 10^{-2} \text{ meters}}{\text{month}}.$$

(b) Hair grows at a rate of 4.3×10^{-9} meters per second. Let x = the amount of time (in seconds) it takes for hair to grow one meter. The following equation must be true.

$$\left(4.3 \times 10^{-9} \frac{\text{meters}}{\text{second}}\right) \times (x \text{ seconds}) = 1 \text{ meter}$$
$$(4.3 \times 10^{-9})(x) = 1$$
$$x = \frac{1}{4.3 \times 10^{-9}}$$
$$x \approx 2.33 \times 10^{8} \text{ seconds.}$$

Therefore, it would take about 2692 days or 7.37 years for hair to grow one meter.

Section 8.2

28. (a) (i) If x is a negative integer, then $|x|$ is a positive integer. Thus, for all negative integers x, $|x| > x$.

(ii) If x is zero or positive, then $|x|$ is zero or positive, respectively. So for all x such that $x \geq 0$, $|x| = x$.

(iii) If x is a negative integer, then $|x| > x$. If x is zero or a positive integer, then $|x| = x$. Integers can only be negative, positive, or zero, so there is no integer such that $|x| < x$.

(iv) For any integer x, either $|x| > x$ or $|x| = x$. Since $|x| \geq x$ means $|x| > x$ or $|x| = x$, we know that $|x| \geq x$ holds for any integer.

(b) $|x| \geq x$ holds for all integers.

Section 8.2

29. Her reasoning is correct. She will apply the theorem which states "If a is an integer, then $a(-1) = -a$." Let $a = -1$. Then because -1 is an integer, we know that $-1(-1) = -(-1)$ by the theorem she applied. Since a negative integer $\times$ a negative integer = a positive integer, we know that $(-1)(-1) = 1$, and thus, $-(-1) = 1$.

Section 8.2

30. Consider debts of $5 and $10. Since it will take more money to pay off the $10 debt, you will be left with *less* money than if you paid off the $5 debt. Alternately, if you put the dollar amounts on the number line, where positive numbers represent assets and negative numbers represent liabilities, then -10 is to the left of -5. Therefore, $-10 < -5$.

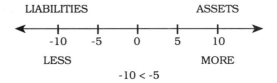

Section 8.2

31. The product of the numbers in row three is 216, so each row, column, and diagonal must contain integers whose product is 216. In column one there is one square empty. The number in this square must be the solution to the equation $-2 \times \square \times 3 = 216$. The missing integer in column one is $216 \div (-6) = -36$. The middle square must contain the solution to the equation $-2 \times \square \times (-18) = 216$. The missing integer in the middle square is $216 \div 36 = 6$. In row two, the missing integer must be the solution to the equation $-36 \times 6 \times \square = 216$. Thus, the missing integer is $216 \div (-216) = -1$. The missing integer in column three must be the solution to the equation $\square \times (-1) \times (-18) = 216$. The missing integer is $216 \div 18 = 12$. The missing integer in row one must be the solution to the equation $(-2) \times \square \times 12 = 216$. The missing integer is $216 \div (-24) = -9$. The multiplicative magic square is shown below.

-2	-9	12
-36	6	-1
3	-4	-18

Section 8.2

32. If x and y are positive integers and $x < y$, then we know that $x \times x < x \times y$. Thus, $x^2 < yx$, and $y \times x < y \times y$, so $yx < y^2$. By the addition approach to less than, $x^2 + c = yx$, for some positive integer c, and $yx + d = y^2$ for some positive integer d. By substitution, $x^2 + c + d = y^2$. Since c and d are positive integers, and the set of integers is closed under addition, $c + d$ is a positive integer. By the addition approach to less than, $x^2 < y^2$.

Section 8.2

33. Keeping in mind that the farmer buys only 100 animals and spends only $1000, consider the maximum number of cows, sheep, and rabbits that can be purchased, and make a table. Since cows cost $50 each, the maximum number of cows that can be purchased is $1000 /$50 = 20 cows.

Cows	Sheep	Rabbits	Total Cost	Total Number
20	0	0	$1000	20
19	0	100	$1000	119
19	1	80	$1000	100

Continue adjusting the number of cows, sheep, and rabbits. The only other solution occurs when the farmer buys only sheep. Since sheep cost $10 each, the maximum number of sheep that can be purchased is $1000/$10 = 100 sheep. Therefore, the farmer buys 19 cows, 1 sheep, and 80 rabbits, or the farmer buys 100 sheep.

Section 8.2

34. Try several examples to see that the statement appears to be true. Prove that if x is a whole number, then x^2 is a multiple of 3 or x^2 is one more than a multiple of 3. Suppose that x is a whole number. Then we know that $x = 3n$, $x = 3n + 1$, or $x = 3n + 2$, for some whole number n. (That is, x is a multiple of 3, one more than a multiple of 3, or two more than a multiple of 3. If x is three more than a multiple of 3, then x is back to being a multiple of 3.) Consider x^2 for each case.
$x^2 = (3n)^2 = 9n^2 = 3(3n^2)$, so x^2 is a multiple of 3.
$x^2 = (3n + 1)^2 = 9n^2 + 6n + 1 = 3(3n^2 + 2n) + 1$, so x^2 is one more than a multiple of 3.
$x^2 = (3n + 2)^2 = 9n^2 + 12n + 4 = 9n^2 + 12n + 3 + 1 = 3(3n^2 + 4n + 1) + 1$, so x^2 is one more than a multiple of 3. Therefore, if x is a whole number, then x^2 is a multiple of 3, or x^2 is one more than a multiple of 3.

Section 8.2

35. Find the cell sums:

	7	11	13
2	9	13	15
3	10	14	16
5	12	16	18

Consider the diagonal sums.
$$9 + 14 + 18 = 41$$
$$12 + 14 + 15 = 41$$

Try another example:

	14	8	21
12	26	20	33
28	42	36	49
9	23	17	30

Consider the diagonal sums.

$$26 + 36 + 30 = 92$$
$$23 + 36 + 33 = 92$$

This procedure appears to produce a square whose diagonal sums are the same. This procedure will always work. Use variables to prove this in general. Let a, b, c, x, y, and z be any numbers.

	x	y	z
a	$a + x$	$a + y$	$a + z$
b	$b + x$	$b + y$	$b + z$
c	$c + x$	$c + y$	$c + z$

Consider the diagonal sums.
$a + x + b + y + c + z$ = sum of the original numbers.
$c + x + b + y + a + z$ = sum of the original numbers.
Since addition is commutative, the diagonal sums are the same.

Section 8.2

36. If $ab = 0$ and $b \neq 0$, then $ab = 0 = 0 \times b$. Since we assume "If $ac = bc$ and $c \neq 0$, then $a = b$.", we can conclude that $a = 0$. Similarly, if $a \neq 0$, then $ab = 0 = a \times 0$, so $b = 0$ by our assumption.

SOLUTIONS - PART A PROBLEMS

Chapter 9: Rational Numbers and Real Numbers, with an Introduction to Algebra

Section 9.1

26. Let n be any nonzero integer. By the definition of equality of rational numbers, to show that two rational numbers are equal, it is necessary to show the "cross products" are equal. Therefore, for $\frac{a}{b}$ to be equal to $\frac{an}{bn}$, we need to show that $a(bn) = b(an)$.

We have $b(an) = (ba)n$ Associativity
$= (ab)n$ Commutativity
$= a(bn)$ Associativity

Therefore, since $b(an) = a(bn)$ we know $\frac{a}{b} = \frac{an}{bn}$.

Section 9.1

27. (a) We need to show that the product of two rational numbers is a rational number. Let $\frac{a}{b}$ and $\frac{c}{d}$ be rational numbers, where b and d are nonzero integers. By the definition of multiplication of rational numbers, $\frac{a}{b} \times \frac{c}{d} = \frac{ac}{bd}$. Since the set of integers is closed under multiplication, ac and bd are integers. Therefore, by the definition of a rational number, $\frac{ac}{bd}$ is a rational number.

(b) We need to show that rational numbers can be multiplied in any order. Let $\frac{a}{b}$ and $\frac{c}{d}$ be rational numbers where b and d are nonzero integers. Then we have

$$\frac{a}{b} \times \frac{c}{d} = \frac{ac}{bd} \qquad \text{Multiplication of rational numbers}$$

$$= \frac{ca}{db} \qquad \text{Commutativity}$$

$$= \frac{c}{d} \times \frac{a}{b} \qquad \text{Multiplication of rational numbers}$$

Therefore, $\frac{a}{b} \times \frac{c}{d} = \frac{c}{d} \times \frac{a}{b}$.

(c) We need to show that rational number multiplication is associative. Let $\frac{a}{b}$, $\frac{c}{d}$, and $\frac{e}{f}$ be rational numbers, where b, d, and f are nonzero integers.

$$\left(\frac{a}{b} \times \frac{c}{d}\right) \times \frac{e}{f} = \left(\frac{ac}{bd}\right) \times \frac{e}{f} \qquad \text{Multiplication of rational}$$
numbers

$$= \frac{(ac)e}{(bd)f} \qquad \text{Multiplication of rational}$$
numbers

$$= \frac{a(ce)}{b(df)} \qquad \text{Associativity for integer}$$
multiplication

$$= \frac{a}{b} \times \left(\frac{ce}{df}\right) \qquad \text{Multiplication of rational}$$
numbers

$$= \frac{a}{b} \times \left(\frac{c}{d} \times \frac{e}{f}\right) \text{Multiplication of rational}$$
numbers

Therefore, $\left(\dfrac{a}{b} \times \dfrac{c}{d}\right) \times \dfrac{e}{f} = \dfrac{a}{b} \times \left(\dfrac{c}{d} \times \dfrac{e}{f}\right)$.

(d) We need to show there exists a rational number so that the product of it and any other rational number is equal to that rational number. Let $\dfrac{a}{b}$ be any rational number where b is a nonzero integer. We know that

$$\frac{a}{b} \times 1 = \frac{a}{b} \times \frac{1}{1}$$

$$= \frac{a \times 1}{b \times 1} \qquad \text{Multiplication of rational}$$
numbers

$$= \frac{a}{b} \qquad \text{Multiplicative identity}$$
for integers

$$1 \times \frac{a}{b} = \frac{1}{1} \times \frac{a}{b}$$

$$= \frac{1 \times a}{1 \times b} \qquad \text{Multiplication of rational}$$
numbers

$$= \frac{a}{b} \qquad \text{Multiplicative identity}$$
for integers

Therefore, since $\dfrac{a}{b} \times 1 = \dfrac{a}{b} = 1 \times \dfrac{a}{b}$, 1 is the identity element of rational-number multiplication.

(e) We need to show that for each rational number there exists another rational number, the inverse, so that the product of the rational number and its inverse is the identity element. Let $\dfrac{a}{b}$ be any nonzero rational number. Then $a \neq 0$ and $b \neq 0$.

$$\frac{a}{b} \times \frac{b}{a} = \frac{ab}{ba} \qquad \text{Multiplication of rational}$$
$$\text{numbers}$$

$$= \frac{ab}{ab} \qquad \text{Commutativity}$$

$$= \frac{a}{a} \times \frac{b}{b} \qquad \text{Multiplication of rational}$$
$$\text{numbers}$$

$$= 1 \times 1 \qquad \text{Since } \frac{m}{m} = 1 \text{ when } m \neq 0$$

$$= 1$$

Therefore, each nonzero rational number $\frac{a}{b}$ has a unique inverse, $\frac{b}{a}$.

Section 9.1

28. (a) $\frac{a}{b} - \frac{c}{d} = \frac{e}{f}$ if, and only if, $\frac{a}{b} = \frac{c}{d} + \frac{e}{f}$.

(b) $\frac{a}{b} - \frac{c}{d} = \frac{a}{b} + \left(-\frac{c}{d}\right)$ by the adding-the-opposite approach. Since $\frac{a}{b} - \frac{c}{d} = \frac{e}{f}$ and $\frac{a}{b} - \frac{c}{d} = \frac{a}{b} + \left(-\frac{c}{d}\right)$, we know $\frac{a}{b} + \left(-\frac{c}{d}\right) = \frac{e}{f}$. Adding $\frac{c}{d}$ to both sides, we get $\frac{a}{b} + \left(-\frac{c}{d}\right) + \frac{c}{d} = \frac{e}{f} + \frac{c}{d}$. Thus, $\frac{a}{b} = \frac{c}{d} + \frac{e}{f}$ by the additive inverse and commutative properties. Therefore, we have established that if $\frac{a}{b} - \frac{c}{d} = \frac{e}{f}$, then $\frac{a}{b} = \frac{c}{d} + \frac{e}{f}$.

Now suppose that $\frac{c}{d} + \frac{e}{f} = \frac{a}{b}$. Adding $-\frac{c}{d}$ to both sides, we get $-\frac{c}{d} + \frac{c}{d} + \frac{e}{f} = -\frac{c}{d} + \frac{a}{b}$. By the additive inverse and commutative properties, we know that $\frac{e}{f} = \frac{a}{b} + \left(-\frac{c}{d}\right)$. Also, $\frac{e}{f} = \frac{a}{b} - \frac{c}{d}$ by the adding-the-opposite approach. Therefore, we have established that if $\frac{c}{d} + \frac{e}{f} = \frac{a}{b}$, then $\frac{e}{f} = \frac{a}{b} - \frac{c}{d}$.

By these two parts together, we have verified the missing-addend approach.

(c) Assuming the missing-addend approach holds, we need to show that $\frac{a}{b} - \frac{c}{d} = \frac{a}{b} + \left(-\frac{c}{d}\right)$. If $\frac{a}{b} - \frac{c}{d} = \frac{e}{f}$, then $\frac{a}{b} = \frac{c}{d} + \frac{e}{f}$ by the missing-addend approach. Adding $-\frac{c}{d}$ to both sides, we have that $\frac{a}{b} + \left(-\frac{c}{d}\right) =$ $\frac{c}{d} + \frac{e}{f} + \left(-\frac{c}{d}\right) = \frac{c}{d} + \left(-\frac{c}{d}\right) + \frac{e}{f}$ (commutative property). Thus, $\frac{a}{b} + \left(-\frac{c}{d}\right) = \frac{e}{f}$ by the additive inverse property. Therefore, by substitution, $\frac{a}{b} - \frac{c}{d} = \frac{a}{b} + \left(-\frac{c}{d}\right)$.

Section 9.1

29. Let $\frac{a}{b}, \frac{c}{d}$, and $\frac{e}{f}$ be rational numbers.

$$\frac{a}{b}\left(\frac{c}{d} + \frac{e}{f}\right) = \frac{a}{b}\left(\frac{cf + de}{df}\right) \qquad \text{Addition of rational numbers}$$

$$= \frac{a(cf + de)}{b(df)} \qquad \text{Multiplication of rational numbers}$$

$$= \frac{a(cf) + a(de)}{b(df)} \qquad \text{Distributive property}$$

$$= \frac{acf + ade}{bdf} \qquad \text{Multiplication of integers}$$

$$= \frac{acf}{bdf} + \frac{ade}{bdf} \qquad \text{Addition of rational numbers}$$

$$= \frac{ac}{bd} \times \frac{f}{f} + \frac{ae}{bf} \times \frac{d}{d} \qquad \text{Multiplication of rational numbers and commutativity}$$

$$= \frac{ac}{bd} + \frac{ae}{bf} \qquad \text{Multiplicative identity for rational numbers}$$

$$= \frac{a}{b} \times \frac{c}{d} + \frac{a}{b} \times \frac{e}{f} \qquad \text{Multiplication of rational numbers}$$

Section 9.1

30. If $\frac{a}{b} < \frac{c}{d}$, then there exists some nonzero rational number $\frac{m}{n}$ so that $\frac{a}{b} + \frac{m}{n} = \frac{c}{d}$. For nonzero rational number e/f, $\frac{a}{b} + \frac{m}{n} + \frac{e}{f} = \frac{c}{d} + \frac{e}{f}$. Then $\frac{a}{b} + \frac{e}{f} + \frac{m}{n} = \frac{c}{d} + \frac{e}{f}$ by the commutative property for rational-number addition. Thus, $\frac{a}{b} + \frac{e}{f} < \frac{c}{d} + \frac{e}{f}$ by the additive approach to less than. Therefore, if $\frac{a}{b} < \frac{c}{d}$ then $\frac{a}{b} + \frac{e}{f} < \frac{c}{d} + \frac{e}{f}$.

Section 9.2

26. Use indirect reasoning. Suppose that there is a rational number $\frac{a}{b}$ such that $\frac{a}{b} = \sqrt{3}$. Then squaring both sides yields $\left(\frac{a}{b}\right)^2 = 3$. So $\frac{a^2}{b^2} = 3$, and $a^2 = 3b^2$. By the Fundamental Theorem of Arithmetic, we know that every whole number has a unique prime factorization. Since perfect squares have an even number of prime factors, a^2 and b^2 have an even number of prime factors. Notice, however, that since 3 is prime, $3b^2$ must have an odd number of prime factors. Since a whole number cannot have an even number *and* an odd number of prime factors, $\sqrt{3}$ must not be rational.

Section 9.2

27. Suppose there is a rational number $\frac{a}{b}$ such that $\frac{a}{b} = \sqrt{9}$. Squaring both sides yields $\left(\frac{a}{b}\right)^2 = 9$. Then $\frac{a^2}{b^2} = 9$ and $a^2 = 9b^2$. By the Fundamental Theorem of Arithmetic, a^2 and $9b^2$ must have the same prime factorization. We know a^2 and b^2 have an even number of prime factors since they are both perfect squares. No contradiction arises now, however, since $9 = 3^2$ and $3^2b^2 = (3b)^2$, which also has an even number of prime factors. Therefore, there is not necessarily a contradiction when trying to show $\sqrt{9}$ is irrational in this way.

Section 9.2

28. Suppose there is a rational number $\frac{a}{b}$ such that $\frac{a}{b} = \sqrt[3]{2}$.

    Cubing both sides yields $\left(\frac{a}{b}\right)^3 = 2$. Then $\frac{a^3}{b^3} = 2$ and $a^3 = 2b^3$. By the Fundamental Theorem of Arithmetic, the prime factorization of a^3 and $2b^3$ must be the same. Note that each prime factor in the prime factorization of a perfect cube has an exponent that is a multiple of 3. If there is a 2 in the prime factorization of a^3 and b^3, then its exponent is a multiple of 3. However, when $2b^3$ is written in its prime factorization, the exponent on the 2 will be one more than a multiple of 3. Since the exponent on the 2 is not the same in the prime factorizations of a^3 and $2b^3$, there is a contradiction. Therefore, $\sqrt[3]{2}$ is irrational.

Section 9.2

29. (a) Suppose there exists a rational number $\frac{a}{b}$ so that $\frac{a}{b} = 5\sqrt{3}$. Multiplying both sides by the rational number $1/5$ yields $\frac{a}{5b} = \sqrt{3}$. Since the set of rational numbers is closed under multiplication, $\frac{a}{5b}$ is a rational number. However, we know $\sqrt{3}$ is irrational. Therefore, since a number cannot be both rational *and* irrational, we have a contradiction. Therefore, $5\sqrt{3}$ is irrational.

    (b) Suppose m is an irrational number and n is a nonzero rational number. We need to show nm is also irrational. Suppose there is a rational number $\frac{a}{b}$ so that $\frac{a}{b} = nm$. Multiplying both sides by the rational number $\frac{1}{n}$ yields $\frac{a}{nb} = m$. Since the set of rational numbers is closed under multiplication, $\frac{a}{nb}$ is rational. However, we have assumed that m is irrational. Since $\frac{a}{nb}$ cannot be both rational *and* irrational, we have a contradiction. Therefore, the product of any nonzero rational number with an irrational number is an irrational number.

Section 9.2

30. (a) Suppose there is a rational number $\frac{a}{b}$ such that

$\frac{a}{b} = 1 + \sqrt{3}$. Subtracting 1 from both sides yields

$\frac{a}{b} - 1 = \sqrt{3}$. Since $\frac{a}{b}$ and 1 are rational numbers,

$\frac{a}{b} - 1$ is a rational number by the closure property.

However, we know $\sqrt{3}$ is irrational. Since a number cannot be both rational and irrational, we have a contradiction. Therefore, $1 + \sqrt{3}$ is irrational.

(b) Let m and n where $n \neq 0$ be rational numbers.

Suppose there is a rational number $\frac{a}{b}$ such that

$\frac{a}{b} = m + n\sqrt{3}$. Subtracting m from both sides yields

$\frac{a}{b} - m = n\sqrt{3}$. Since $\frac{a}{b}$ and m are rational numbers,

$\frac{a}{b} - m = \frac{a - bm}{b}$ is rational by the closure property.

Multiplying both sides by $\frac{1}{n}$ yields $\frac{a - bm}{bn} = \sqrt{3}$. Since

$\frac{1}{n}$ is rational, $\frac{a - bm}{bn}$ is rational. However, we know

that $\sqrt{3}$ is irrational. Therefore, we have a contradiction, so $m + n\sqrt{3}$ must have been irrational.

Section 9.2

31. (a) $6\sqrt{2}$ is irrational since we proved that $\sqrt{2}$ is irrational, and we proved that the product of a rational and an irrational number is irrational from problem 29(b).

(b) $2 + \sqrt{3}$ is irrational since we proved that any number of the form $m + n\sqrt{3}$ is irrational for rational numbers m and n from problem 30(b). (Here $m = 2$ and $n = 1$.)

(c) $5 + 2\sqrt{3}$ is irrational since we proved that any number of the form $m + n\sqrt{3}$ is irrational for rational numbers m and n from problem 30(b). (Here $m = 5$ and $n = 2$.)

Section 9.2

32. If the student checks other examples, he will notice that $\sqrt{a} + \sqrt{b} = \sqrt{a + b}$ only when a or b or both are zero. For example, let $a = 36$ and $b = 64$, then we have

$$\sqrt{36} + \sqrt{64} \overset{?}{=} \sqrt{36 + 64}$$
$$6 + 8 \overset{?}{=} 10$$

$$14 \neq 10.$$

As another example, let $a = 49$ and $b = 0$. Then

$$\sqrt{49} + \sqrt{0} \overset{?}{=} \sqrt{49 + 0}$$
$$7 + 0 \overset{?}{=} 7$$
$$7 = 7.$$

Therefore, $\sqrt{a} + \sqrt{b} \neq \sqrt{a + b}$, except when $a = 0$, $b = 0$, or both a and b are zero.

Section 9.2

33. In this section, $\sqrt{a}$ was defined only when $a \geq 0$. Therefore, the rule $\sqrt{a} \times \sqrt{b} = \sqrt{ab}$ is only true when $a \geq 0$ and $b \geq 0$. It is not true that $\sqrt{-1} \times \sqrt{-1} = \sqrt{(-1)(-1)}$, which is what led to the contradictory statement $-1 = 1$.

Section 9.2

34. Consider multiples of the Pythagorean triple (3, 4, 5), for example (6, 8, 10). This is a Pythagorean triple since $6^2 + 8^2 = 36 + 64 = 100 = 10^2$. Also (9, 12, 15) is a Pythagorean triple since $9^2 + 12^2 = 225 = 15^2$. It is true that $(3n, 4n, 5n)$ is a Pythagorean triple for any counting number n since $(3n)^2 + (4n)^2 = 9n^2 + 16n^2 = 25n^2 = (5n)^2$. Since there are infinitely many counting numbers, there are infinitely many nonzero, whole-number Pythagorean triples. These, however, are not the only Pythagorean triples. Others include (5, 12, 13) and (7, 24, 25) and multiples of these, as well as infinitely many more.

Section 9.2

35. When generating other primitive triples, choose a value for u and then carefully consider the restrictions to find v.

    Let $u = 2$. Choose v such that v is odd, $v < 2$, and v and 2 are relatively prime. Since u and v must also be whole numbers, the only choice for v is 1. Therefore, we have $a = 2(2)(1) = 4$, $b = 2^2 - 1^2 = 4 - 1 = 3$, and $c = 2^2 + 1^2 = 4 + 1 = 5$.

    Let $u = 3$. Choose v such that v is even, $v < 3$, and v and 3 are relatively prime. The only choice for v is 2. Therefore, $a = 2(3)(2) = 12$, $b = 3^2 - 2^2 = 9 - 4 = 5$, and $c = 3^2 + 2^2 = 9 + 4 = 13$.

    Let $u = 8$. Choose v such that v is odd, $v < 8$ and v and 8 are relatively prime. v could be 7, 5, 3, or 1.
    If $v = 7$, then $a = 2(8)(7) = 112$, $b = 8^2 - 7^2 = 15$, and $c = 8^2 + 7^2 = 113$.
    If $v = 5$, then $a = 2(8)(5) = 80$, $b = 8^2 - 5^2 = 39$, and $c = 8^2 + 5^2 = 89$.

If $v = 3$, then $a = 2(8)(3) = 48$, $b = 8^2 - 3^2 = 55$, and $c = 8^2 + 3^2 = 73$.

If $v = 1$, then $a = 2(8)(1) = 16$, $b = 8^2 - 1^2 = 63$, and $c = 8^2 + 1^2 = 65$.

Therefore, five primitive triples are $(5, 12, 13)$, $(15, 112, 113)$, $(39, 80, 89)$, $(48, 55, 73)$, and $(16, 63, 65)$.

Section 9.2
36. Let x = the smallest of the three consecutive integers. Then x, $x + 1$, and $x + 2$ are three consecutive integers. Since adding two of them and dividing by the third is the same as the smallest of the three integers, we can set up three equations and solve to obtain all possible solutions.

Case 1: Add the two smallest integers, x and $x + 1$.
$$(x + x + 1)/(x + 2) = x$$
$$(2x + 1)/(x + 2) = x$$
$$2x + 1 = x^2 + 2x$$
$$0 = x^2 - 1$$
$$0 = (x - 1)(x + 1)$$
$$x = 1 \text{ or } x = -1.$$
Two possible sets of numbers that satisfy the restrictions are 1, 2, 3 and −1, 0, 1.

Case 2: Add the smallest and the largest integers, x and $x + 2$.
$$(x + x + 2)/(x + 1) = x$$
$$(2x + 2)/(x + 1) = x$$
$$2x + 2 = x^2 + x$$
$$0 = x^2 - x - 2$$
$$0 = (x - 2)(x + 1)$$
$$x = 2 \text{ or } x = -1.$$
Only one additional solution is found: 2, 3, 4.

Case 3: Add the two largest integers, $x + 1$ and $x + 2$.
$$(x + 1 + x + 2)/x = x$$
$$(2x + 3)/x = x$$
$$2x + 3 = x^2$$
$$0 = x^2 - 2x - 3$$
$$0 = (x - 3)(x + 1)$$
$$x = 3 \text{ or } x = -1.$$
Only one additional solution is found: 3, 4, 5.

Therefore, the solutions are $(-1, 0, 1)$; $(1, 2, 3)$; $(2, 3, 4)$; or $(3, 4, 5)$.

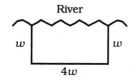

River

w w

$4w$

Section 9.2

37. Consider the diagram to the left:

Let w = width. The rectangle is 4 times as long as it is wide so the length of the region is $4w$. The area of the rectangle is $4w \times w = 4w^2$. The perimeter is $w + w + 4w = 6w$. Since we want to use all 1002 meters of fencing material to enclose the region, we know $6w = 1002$. If $6w = 1002$, then $w = 167$ meters. Therefore, the area is $4w^2 = 4(167)^2 = 111{,}556$ m².

Section 9.2

38. Consider the various forms the rational number and its reciprocal can take. There are three cases to consider.

Case 1: Suppose the rational number and its reciprocal are both integers. We know the sum of two integers is an integer by the closure property of integer addition. The only integers whose reciprocals are also integers are 1 and −1.

Case 2: Suppose the rational number is an integer, n, and its reciprocal is a fraction, $\dfrac{1}{n}$, where $n \neq 1$ and $n \neq -1$. We know the sum of an integer and a non-integer fraction is never an integer.

Case 3: Suppose $\dfrac{a}{b}$ is the rational number in lowest terms (a and b are nonzero and have no common factors). The reciprocal is $\dfrac{b}{a}$. Let's suppose $\dfrac{a}{b} + \dfrac{b}{a}$ is an integer, m. Then

$$\frac{a}{b} + \frac{b}{a} = \frac{a^2 + b^2}{ab} = m \text{ and } abm = a^2 + b^2.$$

Since $a \mid (a^2 + b^2)$ and $a \mid a^2$, we know that

$$a \mid (a^2 + b^2 - a^2) \text{ so } a \mid b^2.$$

This is impossible since a and b have no common factors. Therefore, the sum of a rational number (of this form) and its reciprocal cannot be an integer.

Therefore, there are only two rational numbers, namely 1 and −1, which, when added to their reciprocals, yield integers.

Section 9.2

39.   Consider the diagram

—————— *x* ——————   ——————— *y* ———————

Let *x* and *y* be the lengths of the original pieces of wire. Cut the wire with length *y* at a point that is *m* units from the end. The two pieces have lengths $y - m$ and *m*.

—————— *x* ——————   —— *y – m* —— —— *m* ——

If *m* is the average of the lengths of the other two pieces, then the sum of the lengths of the other two pieces divided by 2 is equal to length *m*.

$$(x + y - m)/2 = m$$
$$x + y - m = 2m$$
$$x + y = 3m$$
$$(x + y)/3 = m.$$

The cut should be made at point *m*, one-third of the way along the wire when they are placed end to end. Since the longer wire is at least $(x + y)/3$ in length, cut the piece of length *m* from it. Therefore, it is possible to cut one of two wires in such a way that one of the three pieces is the average of the lengths of the other two.

Section 9.3

15.   (a) The length of the shadow varies as time passes. It is not the case that the changing time varies as the length of the shadow passes. Therefore, L depends on *n*.   The graph appears to be exponential when compared to the basic function types in this section.

(b) To approximate the function value for each of the given hours past noon, construct a vertical line through the *x*-axis point of interest and note where it intersects the graph. Through that intersection point, construct a horizontal line and note where it intersects the *y*-axis. The *y*-axis intersection point is the function value.

$$L(5) = 150$$
$$L(8) = 1100$$
$$L(2.5) = 50$$

(c) To find the point on the *x*-axis that corresponds to a shadow of 100 meters, construct a horizontal line through the *y*-axis at the given shadow length and note where it intersects the graph. Through that intersection point, construct a vertical line and note where it intersects the *x*-axis. This point on the *x*-axis corresponds to the given shadow length. After $4\frac{1}{2}$ hours, the shadow is 100 meters long. After 6 hours, the shadow is 200 meters long.

(d) The graph stops at *n* = 8 since that is approximately when sundown occurs.

Section 9.3

16. (a) Use a graphing calculator, if available, to sketch the graph of the function $s(t) = -16t^2 + 70t + 55$ over the interval $0 \le t \le 5$.

(b) Consider the graph from part (a). Construct a horizontal line through the *s*-axis at $s(t) = 90$. Notice where this line intersects the graph. Construct a vertical line through each of these intersection points to the *t*-axis. Estimate these values for *t*. They appear to be at about *t* = 0.5 and *t* = 3.5. (Alternatively, use a graphing to "trace" the graph, and identify the times when the ball is about 90 feet above the ground.) Using algebra, we see that

$$90 = -16t^2 + 70t + 55$$
$$0 = -16t^2 + 70t - 35$$

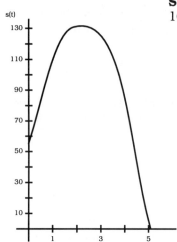

$$t = \frac{-b \pm \sqrt{b^2 - 4ac}}{2a}$$
$$t = \frac{-70 \pm \sqrt{2660}}{-32}$$
$$t \approx 0.58 \text{ or } t \approx 3.8.$$

(c) The ball hits the ground when $s(t) = 0$. We can solve the equation $0 = -16t^2 + 70t + 55$, or we can "trace" the function on the calculator.

$$0 = -16t^2 + 70t + 55$$
$$t = \frac{-70 \pm \sqrt{70^2 - 4(-16)(55)}}{2(-16)}$$
$$t = \frac{-70 \pm \sqrt{8420}}{-32}$$
$$t \approx 5.055.$$

Therefore, the ball hits the ground after about 5.1 seconds.

the *x*-coordinate of the vertex of a parabola can be found by using the formula $\dfrac{-b}{2a}$.

$$\frac{-b}{2a} = \frac{-70}{2(-16)} = \frac{-70}{-32} = 2.1875.$$

Therefore, the ball reaches its maximum height after 2.1875 seconds, at which time its height is $-16(2.1875)^2 + 70(2.1875) + 55 \approx 131.5$ feet.

Section 9.3

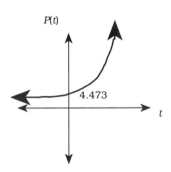

P(t)

4.473

t

17. (a) Use a graphing calculator, if available, to sketch the graph of the function *P*.
    (b) Since *t* represents the number of years since 1980, the value for *t* in 1996 is 16. The population of the world in 1996 can be estimated, by using the function, as
    $$P(16) = 4.473e^{0.01685(16)} \approx (4.473)(1.30944) \approx 5.86$$
    billion people.
    (c) If you have a graphing calculator, then "trace" the graph until you locate the point with a *y*-coordinate of 8. Alternatively, you can locate 8 on the vertical axis of your graph, and then construct a horizontal line through 8. At the point where this line intersects the graph, construct a vertical line, and note where it intersects the *t*-axis. This value, 34.5, is approximately the number of years after 1980 when the world population will reach 8 billion. To get a more accurate value for *t*, you could guess and test *t* values close to the value you just found for *t* in the equation $8 = 4.473e^{0.01685t}$.
    (d) To estimate the current doubling time, choose any point on the graph. Find the point on the graph where the value of P is twice as large. For example, when $t = 0$, there were about $4.473e^{0.01685(0)} = 4.473$ billion people in the world. We need to find *t* when the population is $2(4.473) = 8.946$ billion. This occurs when *t* is about 41. Therefore, the current doubling time is approximately 41 years.

Section 9.3

18. The speed of the cyclist will remain about the same until she gets to the base of the hill. As she begins up the hill, her speed slows (the graph of the function decreases) since it is more difficult to pedal up a hill. At the top of the hill, her speed will stop decreasing (the graph bottoms out). As she begins cycling down the other side, the speed will increase (the graph increases) and will continue to increase so that she will be moving faster than before she started up the hill. Once she reaches the bottom of the

she begins cycling down the other side, the speed will increase (the graph increases) and will continue to increase so that she will be moving faster than before she started up the hill. Once she reaches the bottom of the hill, her speed gradually decreases (the graph decreases) until she resumes travelling at a constant rate on the flat part of the route (the graph is constant as before). Therefore, graph (a) is the best choice.

Section 9.3

19. Graph the function placing n, the number of cricket chirps per minute, on the x-axis, and $T(n)$, the temperature in degrees Fahrenheit, on the y-axis.

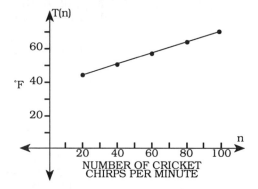

The function appears to be linear. Since it is linear, we can find the formula for $T(n)$ by finding the slope and T - intercept. To find the slope we can use any two points, say $(20,45)$ and $(100,65)$.

$$\text{Slope} = \frac{y_2 - y_1}{x_2 - x_1} = \frac{20}{80} = \frac{1}{4}.$$

Knowing the slope, we can write the function $T(n) = \frac{1}{4} n + b$, where b is the T-intercept. We can use any point to solve for the T-intercept. When $n = 20$, $T(20) = 45$, so $45 = \frac{1}{4} (20) + b = 5 + b$, or $40 = b$.

Therefore, the function is $T(n) = \frac{n}{4} + 40$.

SOLUTIONS - PART A PROBLEMS

Chapter 10: Statistics

Section 10.1

14. (a) We are comparing the number of public and private schools in the United States in 1988. A multiple-bar graph will nicely demonstrate this comparison since for each type of school, the bars which represent the numbers of public and private schools will be side-by-side.

    (b)

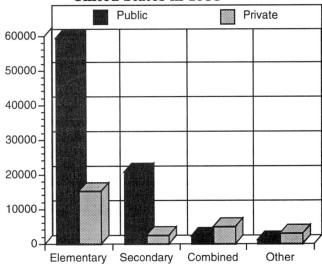

Number of Public and Private Schools in the United States in 1988

Section 10.1

15. (a) In the table, the federal budget revenue for 1992 is divided into parts. In displaying the data, we would like to see how each part compares to the whole budget. A circle graph would allow us to see the relative size of each revenue source. Notice also that dollar amounts are not given - just percentages. Circle graphs generally show relative amounts and not necessarily absolute amounts.

    (b) When constructing the circle graph, recall that the central angle for each sector is found by multiplying the percent by 360°. See the discussion in the text about circle graphs.

A breakdown of the federal budget revenue for 1992 is as follows:

Individual income taxes	47%
Social insurance receipts	31%
Corporate income taxes	13%
Excise taxes	5%
Other sources	4%

The corresponding circle graph is shown below.

Federal Budget Revenue in 1992

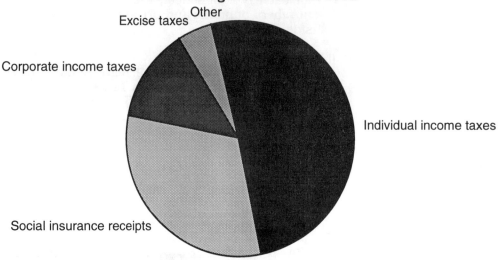

Section 10.1

16. (a) For each year, the data represents the average cost of tuition and fees for public and private U.S. colleges. Notice that many consecutive years are given, so the graph chosen should show any time trends in the data. Line graphs are useful for plotting data over a period of time to indicate trends. A line graph for each school will show the time trends and comparison at the same time.

(b)

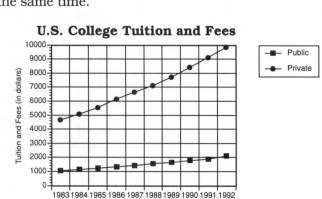

U.S. College Tuition and Fees

(c) Notice that the slope of the line graph of the private college is generally steeper, so over the 10-year period, private school costs increased at a greater rate. We could also consider the percent increase in tuition and fees for each type of school.
Percent increase = [(new cost – old cost)/(old cost)] × 100%.
Percent increase (public) = [(2134 – 1031)/1031] × 100% =106.98% .
Percent increase (private) = [(9841 – 4639)/4639] × 100% = 112.14% .

Therefore, the percent increase for private schools was greater than for public schools over the same time period.

Section 10.1

17. (a) The table lists total pieces of mail for five U.S. cities. A graph should demonstrate how the cities compare to each other and allow for a quick assessment of the total number of pieces of mail for each city. A bar graph or a pictograph would beappropriate. For a pictograph, the quantity to berepresented, pieces of mail, can be drawn easily.

(b)

U.S. Cities Receiving the Most Mail, 1988

Section 10.1

18. (a) The data given is the percent of U.S. households with video cassette recorders over a period of six years. The graph should demonstrate any trend over time and allow for a quick assessment of what fraction of households out of all households have VCR's in the given years. A bar graph is most appropriate in this case.

(b)

U.S. Households with Videocassette Recorders

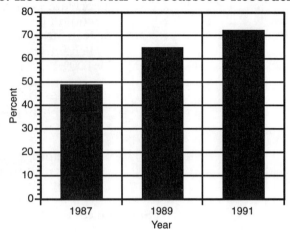

Section 10.1

19. (a) The data given are revenues from three sources over a period of 70 years. We are interested in the trends over time and how the relationship between revenue sources changed over time. A multiple-line graph or a multiple-bar graph will show this relationship.

(b)

Source of School Funding

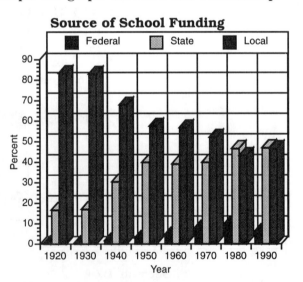

(c) Funds from federal sources increase steadily until sometime during the 1980's then they decrease. State funding nearly doubles by 1940 then increases in small jumps. After providing over 80% of the school funding, local sources decrease steadily, eventually providing less than 50% of the funds.

Section 10.1

20. (a) The graph chosen should allow for a quick visual comparison of percentages in each age group during each of two years. We want the graph to show how the percentage of first births changes with increasing age of mothers. A multiple bar graph can be used to show a trend over the age of the mother.

(b)

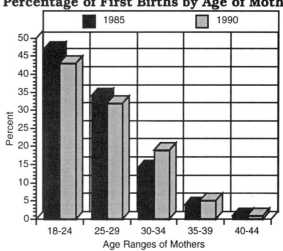

Percentage of First Births by Age of Mother

Section 10.1

21. (a)

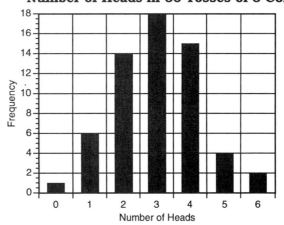

Number of Heads in 60 Tosses of 6 Coin

(b) **Number of Heads in 60 Tosses of 6 Coins**

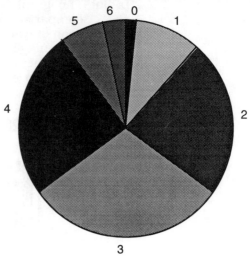

(c) Using the information from the circle graph, we see that, in general, about 10% of the time you can expect to get just one head. If the coin were tossed 100 times, you would expect to get $10\% \times 100 = (0.10)(100) = 10$ heads.

Section 10.1

22. (a) Use a tree diagram to generate all possible outcomes.
All possible outcomes: HH, HT, TH, TT

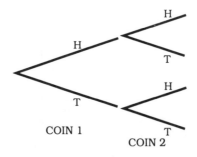

(b) If each outcome is as likely as every other outcome, then each outcome has one chance out of four of occurring. If two coins are tossed 20 times, you would expect both to be heads, HH, 1 out of 4 or 25% of the time. Thus, you would expect to get two heads 25% of 20 times, or $0.25(20) = 5$ times.

(c) Notice that there are two outcomes that correspond to getting exactly one head: HT and TH. Since each of these outcomes has a 1 in 4 chance of occurring, the probability of getting exactly one head is $1/4 + 1/4 =$

2/4 = 1/2. In 20 tosses then, we would expect 1/2 of the outcomes, or 10 outcomes, with exactly one head.

(d) There is only one outcome that corresponds to having exactly two tails: TT. It has a 1 in 4 chance of occurring. Out of 20 tosses, we would expect 1/4 of the outcomes, or 5 outcomes, to have two tails.

(e)

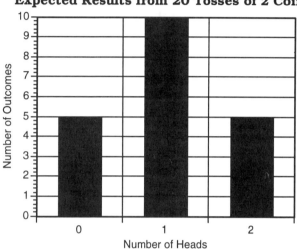

Expected Results from 20 Tosses of 2 Coins

Number of Outcomes (y-axis)

Number of Heads (x-axis)

Section 10.1

23. (a)
$$4^2 + 5^2 + 6^2 \overset{?}{=} 2^2 + 3^2 + 8^2$$
$$16 + 25 + 36 \overset{?}{=} 4 + 9 + 64$$
$$77 = 77, \quad \text{so the equation is true.}$$

$$42^2 + 53^2 + 68^2 \overset{?}{=} 24^2 + 35^2 + 86^2$$
$$1764 + 2809 + 4624 \overset{?}{=} 576 + 1225 + 7396$$
$$9197 = 9197, \quad \text{so the equation is true.}$$

(b) Both equations are true.

(c) Let variables a, b, c, d, e, and f be any whole numbers. Suppose $a^2 + b^2 + c^2 = d^2 + e^2 + f^2$ is a true statement. Then show $(ad)^2 + (be)^2 + (cf)^2 = (da)^2 + (eb)^2 + (fc)^2$ using expanded form.

$$(10a + d)^2 + (10b + e)^2 + (10c + f)^2$$
$$= 100a^2 + 20ad + d^2 + 100b^2 + 20be + e^2 + 100c^2 + 20cf + f^2$$
$$= 100(a^2 + b^2 + c^2) + 20(ad + be + cf) + (d^2 + e^2 + f^2)$$
$$= 100(d^2 + e^2 + f^2) + 20(da + eb + fc) + (a^2 + b^2 + c^2)$$

(substitution and commutative property)

$$= 100d^2 + 100e^2 + 100f^2 + 20da + 20eb +$$
$$20fc + a^2 + b^2 + c^2$$
$$= 100d^2 + 20da + a^2 + 100e^2 + 20eb + b^2 +$$
$$100f^2 + 20fc + c^2$$
$$= (10d + a)^2 + (10e + b)^2 + (10f + c)^2.$$

Therefore, $(ad)^2 + (be)^2 + (cf)^2 = (da)^2 + (eb)^2 + (fc)^2$.

Section 10.1

24. If all four digits are even, then the smallest n^2 can be is 2000 which means the smallest n can be is 46 (since $\sqrt{2000} \approx 44.7$). The largest n^2 can be is 8888, which means the largest n can be is 94 (since $\sqrt{8888} \approx 94.2$). We need only consider possible n's from 46 to 94. Notice that we can eliminate all odd numbers since an odd number multiplied by an odd number is an odd number. Make a list.

Possible n's	n^2	
46	2116	
48	2304	
50	2500	
52	2704	
54	2916	
56	3136	
58	3364	
60	3600	
62	3844	
64	4096	
66	4356	
68	4624	All Even
70	4900	
72	5184	
74	5476	
76	5776	
78	6084	All Even
80	6400	All Even
82	6724	
84	7056	
86	7396	
88	7744	
90	8100	
92	8464	All Even
94	8836	

Therefore, 4624, 6084, 6400, and 8464 are the only four-digit squares whose digits are all even.

Section 10.2

16. Out of a possible 40 points, the class average (mean) was 27.5. The average is found by adding up the total points for all girls and boys and dividing by the number of students. (19 + 11 = 30.)

$$\frac{\text{Total Score For Girls} + \text{Total Score For Boys}}{30} = 27.5.$$

Since the total score for girls is 532 we can substitute that value into the equation.

$$\frac{532 + \text{Total Score for Boys}}{30} = 27.5$$

$$532 + \text{Total Score for Boys} = 825$$

$$\text{Total Score for Boys} = 293.$$

Notice that 40 is the total number of points possible on the test and is unnecessary information.

Section 10.2

17. Recall how the average score was found. After the scores are totaled, the total is divided by 100. Two more students take the test. The new average is found by adding the two new scores to the original total and dividing by 102.

Original Average: $\dfrac{\text{Total of 100 Scores}}{100} = 77.1$

Total of 100 Scores = 7710

New Average: $\dfrac{\text{Total of 100 Scores} + \text{Two New Scores}}{102} =$

$$\frac{7710 + 125}{102} = 76.81$$

Therefore, the new average is 76.81.

Section 10.2

18. The mean score of 41.6 was found by dividing the sum of the math scores by 35. Set up an equation and solve for the unknown value.

$$\frac{\text{Sum of Scores}}{35} = 41.6$$

$$\text{Sum of Scores} = (41.6)(35) = 1456.$$

Notice that the information about the standard deviation is unnecessary.

Section 10.2

19. (a) The median score is the "middle score" or the "halfway" point in a list of scores that is arranged in increasing or decreasing order. Half means 50%, so the median score is the 50th percentile.

    (b) Consider a normal distribution and notice where the z-scores −2, −1, 1, and 2 fall. For a standard normal distribution, the standard deviation is 1. Every normal distribution has about 68% of the distribution within 1 standard deviation of the mean and about 95% of the distribution within 2 standard deviations of the mean. Therefore, for the standard normal distribution, about 68% of the distribution falls between −1 and 1, and about 95% of the distribution falls between −2 and 2.

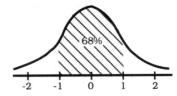 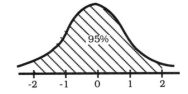

Because of symmetry in all normal distributions, about 34% of the distribution falls between −1 and 0 or between 0 and 1. About 47.5% of the distribution falls between −2 and 0 or 0 and 2.

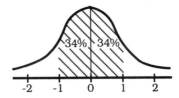 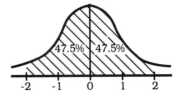

Since 50% of the distribution falls below 0, 50% + 34% = 84% falls below 1 and 50% + 47.5% = 97.5% falls below 2.

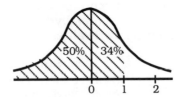 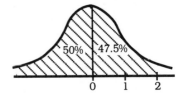

Therefore, the 84th percentile has a z-score of 1 and the 97.5th percentile has a z-score of 2.

We know that 50% of the distribution falls below zero and 34% falls between −1 and 0. Therefore, 50% −

34% = 16% falls below −1. The 16th percentile has a z-score of −1.

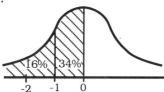

Since 47.5% of the distribution falls between −2 and 0, 50% − 47.5% = 2.5% falls below −2. Thus, the 2.5th percentile has a z-score of −2.

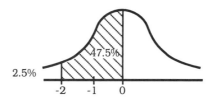

(c) The 50th percentile is the median. In the normal distribution, the median and the mean are equal. Therefore, since the mean of the distribution is 65, the 50th percentile is 65.
The 16th percentile is the point that is 1 standard deviation below the mean (where z = −1). Therefore, 65 − 10 = 55, and the 16th percentile is 55.
The 84th percentile is the point that is 1 standard deviation above the mean (where z = 1). Therefore, 65 + 10 = 75, and the 84th percentile is 75.
Since 99.7% of the distribution is within 3 standard deviations of the mean, 49.85% of the distribution falls between the mean and 3 standard deviations above the mean. Therefore, 50% + 49.85% = 99.85% of the distribution falls below z = 3. The 99.85th percentile is 65 + 10 + 10 + 10 = 95.

Section 10.2

20. We want to find Lora's place relative to the rest of the class. We need to find on which test she has the better z-score. On test 1, the mean score was 81, and the standard deviation was 6.137. Her z-score for test 1 is

$$z = \frac{85 - 81}{6.137} \approx 0.65.$$

(Note: Remember that $z = \frac{\text{mean score} - \text{Lora's score}}{\text{standard deviation}}$.)

On test 2, the mean score was 80.33, and the standard deviation was 13.732. Thus, Lora's z-score for test 2 is

$$z = \frac{89 - 80.33}{13.732} \approx 0.63.$$

Since her z-score was higher on test 1, and a positive z-score is a score above the mean, Lora did better relative to the rest of the class on test 1.

Section 10.2
21. (a) Two normal distributions with different variances will have different spreads. Both distributions will still be bell shaped. If they have the same means, then they will be centered at the same place. The distribution with the smaller variance will have data that is more concentrated around the mean forcing it to have a higher "peak".

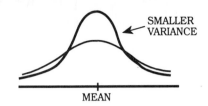

(b) Two normal distributions with different means will be centered in different places. Since the variances are the same, the shape will be the same. The distribution with the larger mean will be shifted to the right.

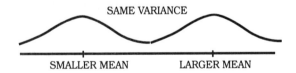

Section 10.2
22. Recall that the z-score of a number indicates how many standard deviations the number is away from the mean.
    (a) $z = (69.2 - 60.3) / 7.82 = 1.14$.
    (b) $z = (75.9 - 60.3) / 7.82 = 1.99$.
    (c) The distribution of scores is normal. The z-score for Miss Brown's class is approximately 2. In the standard normal distribution, we know that about 95% of the data falls between −2 and 2. Therefore about 47.5% of data falls between 0 and 2. Since 50% of the data falls below 0, 50% + 47.5% = 97.5% of the scores fall below Miss Brown's average.

Section 10.2
23. (a) Using the fact that area = width × height, find the total area under the histogram. Notice that the width in each case is 10. Total area = 10(0.1) + 10(0.1) + 10(0.4) + 10(0.1) + 10(0.3) = 1 + 1 + 4 + 1 + 3. So, the total area = 10. Since the total area is 10 square units, the median will occur at the point that leaves 5

square units of area to the right and 5 square units of area to the left. The rectangles above the 10 and 20 each contain 1 square unit of area. Since the rectangle above the 30 contains 4 square units, and we only need 3 more square units of area, the vertical line will fall $\frac{3}{4}$ of the way into the box above the 30. The width of the box is 10 units, and $\frac{3}{4}(10) = 7.5$. Since the rectangle starts at 25, add 7.5, and draw the vertical line at the point 25 + 7.5 = 32.5. This is the median.

(b) Notice that the data points are 10's, 20's, 30's, 40's, and 50's. We do not have a list of each data point. We only know the relative frequency. If there were 10 scores, then there would be 0.1(10) =1 of each of the values 10, 20, and 40. There would be 0.4(10) = 4 of the data value 30. There would be 0.3(10) = 3 of the data value 50. Ordering the data points gives 10, 20, 30, 30, 30, 30, 40, 50, 50, 50. Choosing the middle point gives a median of 30. The two medians are not the same.

Section 10.2

24. Since he must circle the track at an average speed of 60 mph, he will travel 10 miles at 60 mph. We know that distance = rate × time, or time = distance/rate. At 60 mph, he has $t = 10/60 = 1/6$ hours, or 10 minutes, to finish his laps. During the first lap he averaged 30 mph for 5 miles. Therefore, it took $t = 5/30 = 1/6$ hours, or 10 minutes, to complete the first lap. It is impossible for him to qualify since he has no time left in which to complete the second lap.

Section 10.2

25. In a 3 × 3 × 3 cube, there are 3 tiers of 9 cubes each. Altogether this solid collection of cubes has 6 faces: top, bottom, and 4 sides.

(a) Any 1 × 1 × 1 cube which has no faces painted has to lie entirely inside the collection of cubes. Consider the top and bottom tiers of 1 × 1 × 1 cubes. Each 1 × 1 × 1 cube has at least one face showing and, therefore, painted. In the middle tier of the 1 × 1 × 1 cubes, only the one center 1 × 1 × 1 cube has no faces painted, as shown next.

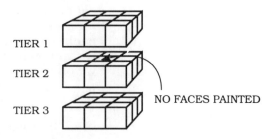

TIER 1

TIER 2

TIER 3   NO FACES PAINTED

Any $1 \times 1 \times 1$ cube that has only one face painted could not be on an edge of the $3 \times 3 \times 3$ cube. Consider each side of the $3 \times 3 \times 3$ cube. There is only one $1 \times 1 \times 1$ cube on each side which does not lie on an edge. Since there are 6 sides of the $3 \times 3 \times 3$ cube, there are $6(1) = 6$, or six, $1 \times 1 \times 1$ cubes with only one face painted.

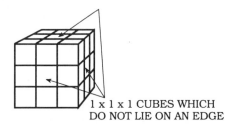

$1 \times 1 \times 1$ CUBES WHICH
DO NOT LIE ON AN EDGE

Any $1 \times 1 \times 1$ cube with three faces painted would have to be a corner cube. In any cube, there are eight corners, so there are eight $1 \times 1 \times 1$ cubes with three faces painted.

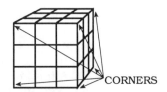

CORNERS

In the $3 \times 3 \times 3$ cube, there are $3(9) = 27$, or twenty-seven, $1 \times 1 \times 1$ cubes. Since 8 of them have 3 faces painted, 6 have 1 face painted, and 1 has no faces painted, $27 - 8 - 6 - 1 = 12$ cubes must have 2 faces painted,

(b) A $4 \times 4 \times 4$ cube has 4 tiers of 16 cubes. Any $1 \times 1 \times 1$ cube with no faces painted would lie in the middle two tiers of cubes since any cube in the top or bottom tiers would have at least one face painted. A single tier of sixteen $1 \times 1 \times 1$ cubes has 4 interior cubes surrounded by 12 edge cubes.

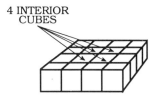

4 INTERIOR CUBES

Since there are two tiers such that the 4 interior cubes have no faces painted, there are 2(4) = 8, or eight, $1 \times 1 \times 1$ cubes with no faces painted. Any $1 \times 1 \times 1$ cube with only 1 face painted will not be on an edge. Consider one side of the $4 \times 4 \times 4$ cube. There are four $1 \times 1 \times 1$ cubes that do not lie on an edge. Since there are 6 faces on any cube, there are 6(4) = 24, or twenty-four, $1 \times 1 \times 1$ cubes with 1 face painted.

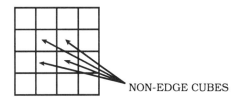

NON-EDGE CUBES

Any $1 \times 1 \times 1$ cube with 3 faces painted would have to be a corner cube. Since there are eight corners on any cube, there are eight $1 \times 1 \times 1$ cubes with 3 faces painted. In the $4 \times 4 \times 4$ cube there are 4(16) = 64, or sixty-four, $1 \times 1 \times 1$ cubes. Since 8 have 3 faces painted, 24 have 1 face painted, and 8 have no faces painted, 64 − 8 − 24 − 8 = 24 have 2 faces painted. Consider a $5 \times 5 \times 5$ cube. It has 5 tiers of 25 cubes. Any $1 \times 1 \times 1$ cube with no faces painted will be in the middle 3 tiers of cubes. A single tier of twenty-five $1 \times 1 \times 1$ cubes has interior cubes surrounded by sixteen $1 \times 1 \times 1$ edge cubes. There are 3(9) = 27, or twenty-seven, $1 \times 1 \times 1$ cubes with no faces painted. Any $1 \times 1 \times 1$ cube with one face painted will not be on an edge. Consider one side of the $5 \times 5 \times 5$ cube. There are nine $1 \times 1 \times 1$ cubes that do not lie on an edge. Since there are six faces on any cube, there are 6(9) = 54, or fifty-four, $1 \times 1 \times 1$ cubes with one face painted. Any $1 \times 1 \times 1$ cube with three faces painted would have to be a corner cube. There are eight corners on any cube, so there are eight $1 \times 1 \times 1$ cubes with three faces painted. In the $5 \times 5 \times 5$ cube, there are 5(25) = 125, or one-hundred twenty-five, $1 \times 1 \times 1$ cubes. Since 8 have 3 faces painted, 54 have 1 face painted, and 27 have no faces painted, 125 − 8 − 54 − 27 = 36 have 2 faces painted.

(c) Make a table and look for a pattern. Let n = a whole number.

Cube	n	no faces painted	1 face painted	2 faces painted	3 faces painted	4 or more faces painted
$3 \times 3 \times 3$	3	$1 = 1^3$	$6 = 6 \times 1^2$	$12 = 12 \times 1$	8	0
$4 \times 4 \times 4$	4	$8 = 2^3$	$24 = 6 \times 2^2$	$24 = 12 \times 2$	8	0
$5 \times 5 \times 5$	5	$27 = 3^3$	$54 = 6 \times 3^2$	$36 = 12 \times 3$	8	0
$n \times n \times n$	n	$(n-2)^3$	$6(n-2)^3$	$12(n-2)$	8	0

Notice there can be no $1 \times 1 \times 1$ cubes that have 4 or more faces painted. Each cube touches at least 3 other cubes. Any cube with 4 or more faces painted would have to have at least 4 sides showing.

SOLUTIONS - PART A PROBLEMS

Chapter 11: Probability

Section 11.1

18. (a) Use ordered pairs to represent the outcomes. The first number in the ordered pair represents the number of spots on the first die. The second number represents the number of spots on the second die. The sum for each pair is indicated.

(1,1) = 2	(1,2) = 3	(1,3) = 4	(1,4) = 5	(1,5) = 6	(1,6) = 7
(2,1) = 3	(2,2) = 4	(2,3) = 5	(2,4) = 6	(2,5) = 7	(2,6) = 8
(3,1) = 4	(3,2) = 5	(3,3) = 6	(3,4) = 7	(3,5) = 8	(3,6) = 9
(4,1) = 5	(4,2) = 6	(4,3) = 7	(4,4) = 8	(4,5) = 9	(4,6) = 10
(5,1) = 6	(5,2) = 7	(5,3) = 8	(5,4) = 9	(5,5) = 10	(5,6) = 11
(6,1) = 7	(6,2) = 8	(6,3) = 9	(6,4) = 10	(6,5) = 11	(6,6) = 12

Sum	2	3	4	5	6	7	8	9	10	11	12
Ways	1	2	3	4	5	6	5	4	3	2	1

(b) A: The sum is prime. The prime numbers up to 12 are 2, 3, 5, 7, and 11. P(A) = P(2) + P(3) + P(5) + P(7) + P(11). Notice that there are 36 possible sums, so, for example, since there are 2 ways to obtain a sum of 3, the probability of obtaining a 3 is 2/36. Therefore, P(A) = 1/36 + 2/36 + 4/36 + 6/36 + 2/36 = 15/36 = 5/12.

B: The sum is a divisor of 12. The divisors of 12 are 1, 2, 3, 4, 6, and 12. Of these, the only ones that are a possible sum of two dice are 2, 3, 4, 6, and 12. Thus, we have P(B) = P(2) + P(3) + P(4) + P(6) + P(12) = 1/36 + 2/36 + 3/36 + 5/36 + 1/36 = 12/36 = 1/3.

C: The sum is a power of 2. The powers of 2 that are also possible sums are 2, 4, and 8. Thus, P(C) = P(2) + P(4) + P(8) = 1/36 + 3/36 + 5/36 = 9/36 = 1/4.

D: The sum is greater than 3. There are many sums that are greater than 3. It may be easier to consider the complement of event D, or $\overline{D}$ The sum is less than or equal to 3. So, $\overline{D}$ = {sum = 2,3}, and P($\overline{D}$) = P(2) + P(3) = 1/36 + 2/36 = 3/36 = 1/12. Since P(D) = 1 – P($\overline{D}$), P(D) = 1 – 1/12 = 11/12. Therefore, the probability of obtaining a sum greater than 3 is 11/12.

Section 11.1

19. (a) $m(S)$ = 100 miles, the distance from Albany to Binghamton, as indicated in the next diagram.

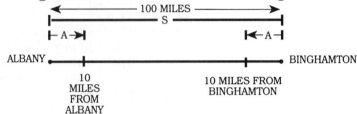

   (b) Since event A involves the part of the road that is 10 miles from either city, $m(A)$ = 20 miles.

   (c) $P(A) = \dfrac{m(A)}{m(S)} = \dfrac{20}{100} = \dfrac{1}{5}$.

Section 11.1

20. Refer to problem 19. The probability that the dart hits the bull's-eye is determined by comparing the area of the bull's-eye to the area of the entire dart board. A is the event the dart hits the bull's-eye. $m(A) = \pi(1)^2 = \pi$ square units. S is the sample space, so S is the whole dart board. Since the radius of the dart board is 4 units, $m(S) = \pi(4)^2 = 16\pi$ square units. Therefore, $P(A) = \dfrac{m(A)}{m(S)} = \dfrac{\pi}{16\pi} = \dfrac{1}{16}$. The probability that the dart hits the bull's-eye is $\dfrac{1}{16}$.

Section 11.2

12. The digits are 0, 1, 2, 3, 4, 5, 6, 7, 8, 9.
   (a) Since a digit can be repeated, there are 10 possible choices for each numeral in the code. Hence, there are $10 \times 10 = 100$ possible two-digit code numbers.
   (b) There is a restriction on the first numeral in the code. The digit 0 cannot be used, so there are only 9 possible choices. Since there are no restrictions on the other two numerals, they each can be chosen in 10 ways. Thus, there are $9 \times 10 \times 10 = 900$ possible 3-digit identification code numbers.
   (c) Notice that no digit can be repeated. There are 10 choices for the first numeral. The second numeral cannot be a repeat of the first, so there are only 9 choices for it. The third numeral can be 1 of 8 numerals since it cannot be a repeat of either of the first two. Finally, there are only 7 possible choices for the fourth numeral, since 3 digits have been used. Therefore, there are $10 \times 9 \times 8 \times 7 = 5040$ possible bicycle lock numbers.
   (d) The only restriction is that the first digit cannot be zero. Consequently, the first digit can be 1 of 9 numerals. There are no restrictions on the rest of the

digits, so there are 10 choices for each of the four remaining digits. Thus, there are 9 × 10 × 10 × 10 × 10 = 90,000 possible five-digit zip code numbers.

Section 11.2

13. (a) There are 3 finishing places to fill. The first finishing place can be filled in 8 ways since there are 8 horses. Once a horse has finished in a certain spot, he cannot finish in another spot, so repetition is not allowed. The second finishing place can be filled in 7 ways. The third finishing place can be filled in 6 ways. Hence, there are 8 × 7 × 6 = 336 finishing orders.

(b) We are still considering 3 finishing places. This time, however, there are only 3 horses to consider. Therefore, there are 3 × 2 × 1 = 6 finishing orders.

(c) Determining the probability that these three particular horses are the top finishers involves comparing the number of ways Lucky One, Lucky Two, and Lucky Three can finish in the top three places to the number of ways that the top three places can be determined in an eight-horse race. Compare part (b) to part (a). P(A) = 6/336 = 1/56.

Section 11.2

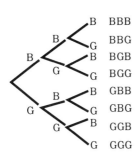

14. (a) There are eight possible orders of boys and girls.

(b) The number of outcomes involving...

All girls,	1: GGG
Two girls, one boy,	3: BGG, GBG, GGB
One girl, two boys,	3: BBG, BGB, GBB
No girls,	1: BBB

(c) This result is the same as the third row of Pascal's Triangle, if we look at the number of outcomes for just the girls (or boys).

All Girls	Two Girls	One Girl	No Girls
1	3	3	1

Section 11.2

15. (a) The number of ways to hit the target when shooting four times should be the 4th row of Pascal's Triangle.

Number of hits:	4	3	2	1	0
Number of ways:	1	4	6	4	1
Probability:	1/16	4/16	6/16	4/16	1/16

Each trial will result in a hit or miss (2 outcomes). Therefore, there are 2 × 2 × 2 × 2 = 16 possible sequences of hits or misses when shooting four times.

(b) Compare the probability for 3 hits out of 3 shots with the probability for 3 hits and 1 miss out of 4 shots.

P(3 hits) = 1/8   (from the table in the book)
P(3 hits and 1 miss) = 4/16 = 1/4   (from our table)

Thus, it is twice as likely that you will get 3 hits and 1 miss in 4 shots as 3 hits out of 3 shots.

Section 11.2

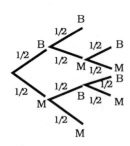

16. (a) No matter how many games are won or lost, the probability that either team wins any game is still 1/2.

(b) Find the probability of event BB. Each game is independent of every other game, so we multiply probabilities to get P(BB) = $\frac{1}{2} \times \frac{1}{2} = \frac{1}{4}$. If Milwaukee wins the series after losing the first game, then Milwaukee wins the final two games. So, P(BMM) = $\frac{1}{2} \times \frac{1}{2} \times \frac{1}{2} = \frac{1}{8}$.

(c) (i) When three games are played, there are two ways for Boston to win: BMB or MBB. We know that P(BMB) = 1/8 and P(MBB) = 1/8. Since Boston can win the series if either event occurs, and the events are mutually exclusive, we can add the probabilities to get P(BMB or MBB) = P(BMB) + P(MBB) = $\frac{1}{8} + \frac{1}{8} = \frac{1}{4}$.

(ii) In this best two out of three series, there are three ways for Milwaukee to win: BMM, MBM, or MM. Since the events are mutually exclusive, we add the probabilities to get P(Milwaukee wins) = P(BMM) + P(MBM) + P(MM) = $\frac{1}{8} + \frac{1}{8} + \frac{1}{4} = \frac{1}{2}$.

(iii) There are four ways for the series to require three games to decide a winner: BMB, BMM, MBB, or MBM. We add the probabilities to get P(3 games required) = $\frac{1}{8} + \frac{1}{8} + \frac{1}{8} + \frac{1}{8} = \frac{1}{2}$.

Section 11.2

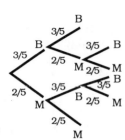

17. (a) The probability that Boston or Milwaukee wins is 1. The probability the Celtics win an individual game is 3/5. The probability that Milwaukee wins an individual game is $1 - \frac{3}{5} = \frac{2}{5}$.

(b) See the tree to the left.

(c) P(MM) = $\frac{2}{5} \times \frac{2}{5} = \frac{4}{25}$.

(d) The probability that Boston wins the series after losing the second game involves the outcome BMB. We have

P(BMB) = $\frac{3}{5} \times \frac{2}{5} \times \frac{3}{5} = \frac{18}{125}$.

(e) The probability that the series goes for three games involves four outcomes. Since there are two outcomes that correspond to the event not happening, consider the probability of the complement of the event, that is, the probability that the series ends in 2 games. The complement involves outcomes BB and MM. We find that P(BB or MM) = P(BB) + P(MM) = $\frac{3}{5} \times \frac{3}{5} + \frac{2}{5} \times \frac{2}{5} = \frac{9}{25} + \frac{4}{25} = \frac{13}{25}$. Therefore, the probability that the series goes for 3 games = 1 – the probability that the series goes for two games = $1 - \frac{13}{25} = \frac{12}{25}$.

(f) The probability that Boston wins the series involves the outcomes BB, BMB, and MBB.

$$P(\text{Boston wins}) = P(\text{BB or BMB or MBB})$$
$$= P(\text{BB}) + P(\text{BMB}) + P(\text{MBB})$$
$$= \frac{3}{5} \times \frac{3}{5} + \frac{3}{5} \times \frac{2}{5} \times \frac{3}{5} + \frac{2}{5} \times \frac{3}{5} \times \frac{3}{5}$$
$$= \frac{9}{25} + \frac{18}{125} + \frac{18}{125}$$
$$= \frac{81}{125}.$$

Section 11.2

18. (a) Since the probability that team A wins any game is $\frac{2}{3}$, label each branch of the tree $\frac{2}{3}$. The probability that team A wins the series in 3 games is P(AAA) = $\frac{2}{3} \times \frac{2}{3} \times \frac{2}{3} = \frac{8}{27}$.

(b) If team A wins with probability $\frac{2}{3}$, then team B wins with probability $1 - \frac{2}{3} = \frac{1}{3}$.

$$P(\text{BAAA}) = \frac{1}{3} \times \frac{2}{3} \times \frac{2}{3} \times \frac{2}{3} = \frac{8}{81} \qquad P(\text{ABAA}) = \frac{2}{3} \times \frac{1}{3} \times \frac{2}{3} \times \frac{2}{3} = \frac{8}{81}$$

The probability of each event is $\frac{8}{81}$, so it does not matter which of the first three games A loses. On each path, the same probabilities are rearranged. The probability that team A wins the series in four games is P(BAAA or ABAA or AABA) = $3 \times \frac{8}{81} = \frac{8}{27}$.

(c) There are six ways for team A to win in a five-game series: AABBA, ABABA, BAABA, ABBAA, BABAA, and BBAAA. Notice that it will not matter which two of the first four games team A loses. Since the probabilities are the same (just rearranged), we need to calculate only one of the probabilities:

$$P(AABBA) = \frac{2}{3} \times \frac{2}{3} \times \frac{1}{3} \times \frac{1}{3} \times \frac{2}{3} = \frac{8}{243}.$$

Therefore, the probability that team A wins the series in five games is $6 \times \frac{8}{243} = \frac{48}{243} = \frac{16}{81}$.

(d) The probability that A wins the series corresponds to the sum: probability A wins in 3 games + probability A wins in 4 games + probability A wins in 5 games. Therefore, from parts (a), (b), and (c) we know that P(A wins) $= \frac{8}{27} + \frac{8}{27} + \frac{16}{81} = \frac{64}{81}$. The probability that B wins is the complementary event. So, we know that P(B wins) $= 1 - \frac{64}{81} = \frac{17}{81}$.

Section 11.2

19. (a) See the tree to the left.

(b) There are 3 questions with 2 choices each. Thus, there are $2 \times 2 \times 2 = 8$ outcomes. There are 8 possible outcomes that correspond to the number of end branches in the tree diagram.

(c) Each sequence of answers is different, so only one can be correct. The single outcome that corresponds to all correct answers is TTT.

(d) The probability of guessing all the correct answers is

$$P(TTT) = \frac{1}{2} \times \frac{1}{2} \times \frac{1}{2} = \frac{1}{8}.$$

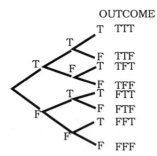

OUTCOME
T TTT
F TTF
T TFT
F TFF
T FTT
F FTF
T FFT
F FFF

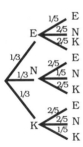

Section 11.2

20. (a) The tree diagram will have two stages corresponding to the two sock selections. Each time you reach into the drawer, there are three possible colors that could be drawn. Since there are two of each color, the probability of getting any one color (blue, brown, or black) is $\frac{2}{6}$ or $\frac{1}{3}$. Once a sock is selected, the possibilities change for the next selection since fewer socks are in the drawer. For example, suppose that you select a blue sock first. On the second draw, there are two black and two brown socks, but one blue sock. Therefore, on the second draw, the probability of obtaining a black sock is $\frac{2}{5}$, a brown $\frac{2}{5}$, and a blue $\frac{1}{5}$.

    (b) The sample space is the set {EE, EN, EK, NE, NN, NK, KE, KN, KK}.

    (c) A matched pair means two of the same color sock. Three outcomes correspond to the event that you get a matched pair: {EE, NN, KK}.

    (d) P(EE or NN or KK) = P(EE) + P(NN) + P(KK)
    $$= \frac{1}{3} \times \frac{1}{5} + \frac{1}{3} \times \frac{1}{5} + \frac{1}{3} \times \frac{1}{5}$$
    $$= \frac{1}{15} + \frac{1}{15} + \frac{1}{15}$$
    $$= \frac{1}{5}.$$

    (e) The probability of getting a matched blue pair, EE, is
    $$P(EE) = \frac{1}{3} \times \frac{1}{5} = \frac{1}{15}.$$

Section 11.2

21. Since there are three different colors, in the worst case you could draw one of each color before obtaining a matched pair. The fourth draw will always match one of the socks you already have. Therefore, the minimum number of socks you nccd to take to be sure you have a matched pair is four.

Section 11.2

22. Since there are three puppies, each of which can be male or female, the only possibilities for the sexes of the puppies are MMM, MMF, MFM, FMM, MFF, FMF, FFM, FFF. Therefore, there are eight elements in the sample space. Knowing that one of the puppies is a female eliminates the MMM element, leaving seven elements in the sample space. The probability that either of the remaining puppies is male involves six out of the seven remaining elements: MMF, MFM, FMM, MFF, FMF, and FFM. Therefore, the probability is $\frac{6}{7}$.

Section 11.2
23.

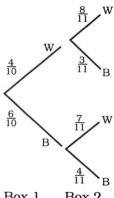

Box 1   Box 2

In box 1, there are 10 balls: 4 white and 6 black. The probability of choosing a white ball is $\frac{4}{10}$, of choosing a black $\frac{6}{10}$. If we pick a white ball and place it in box 2, then we will have increased the total number of balls to 11. In box 2 then, there would be 8 white balls and 3 black balls. Thus, the probability of choosing a white ball is $\frac{8}{11}$, of choosing a black $\frac{3}{11}$.

If, on the other hand, a black ball is chosen from box 1 and placed in box 2, then the number of balls still increases to 11, but the number of white balls remains 7, and the number of black balls increases to 4. In this case, the probability of choosing a white ball from box 2 is $\frac{7}{11}$, of choosing a black $\frac{4}{11}$. A white ball can be chosen from the second box in two ways: (1) choose a black ball from box 1 and choose a white ball from box 2 *or* (2) choose a white ball from box 1 and choose a white ball from box 2. Thus, the probability of choosing a white ball from box 2 is

$$\frac{6}{10} \times \frac{7}{11} + \frac{4}{10} \times \frac{8}{11} = \frac{42}{110} + \frac{32}{110} = \frac{37}{55} \approx 0.673.$$

Section 11.3

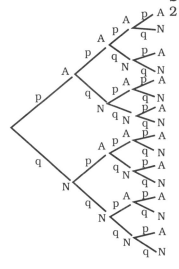

A 21. (a) When the teams are closely matched, it is not likely that one team will "run away" with the series. The prospects for a long series would increase when the teams are closely matched.

(b) Construct a tree diagram. If the probability that the American League team wins any game is p, then the probability that the National League team wins any game is $1 - p = q$. The tree diagram will have four stages corresponding to each game. (See the tree to the left.) Consider the event AAAA. We see that P(AAAA) = P(American League wins in 4 games) = $p \times p \times p \times p = p^4$.

(c) Consider the diagram from part (b). Consider event NNNN. We see that P(NNNN) = P(National League team wins in 4 games) = $q \times q \times q \times q = q^4$.

(d) For the series to end in four games, either the American League team would have to win in four games or the National League team would have to win in four games. These are mutually exclusive events. The probability that the series ends in four games is the sum of the probability that the American League team wins in four games plus the probability that the National League team wins in four games = $p^4 + q^4$.

(e) Given the odds in favor of an event, recall how the probability is calculated. If the odds in favor of an event are $a{:}b$, then the probability of the event occurring is $\dfrac{a}{a + b}$.

Odds favoring National League	1:1	2:1	3:1	3:2
p	$\dfrac{1}{1 + 1} = \dfrac{1}{2}$	$\dfrac{2}{2 + 1} = \dfrac{2}{3}$	$\dfrac{3}{3 + 1} = \dfrac{3}{4}$	$\dfrac{3}{3 + 2} = \dfrac{3}{5}$
$q = 1 - p$	$1 - \dfrac{1}{2} = \dfrac{1}{2}$	$1 - \dfrac{2}{3} = \dfrac{1}{3}$	$1 - \dfrac{3}{4} = \dfrac{1}{4}$	$1 - \dfrac{3}{5} = \dfrac{2}{5}$
P(Amer. in 4)	$\left(\dfrac{1}{2}\right)^4 = \dfrac{1}{16}$	$\left(\dfrac{2}{3}\right)^4 = \dfrac{16}{81}$	$\left(\dfrac{3}{4}\right)^4 = \dfrac{81}{256}$	$\left(\dfrac{3}{5}\right)^4 = \dfrac{81}{625}$
P(Natl. in 4)	$\left(\dfrac{1}{2}\right)^4 = \dfrac{1}{16}$	$\left(\dfrac{1}{3}\right)^4 = \dfrac{1}{81}$	$\left(\dfrac{1}{4}\right)^4 = \dfrac{1}{256}$	$\left(\dfrac{2}{5}\right)^4 = \dfrac{16}{625}$
P(4-game series)	$\dfrac{1}{16} + \dfrac{1}{16} = \dfrac{1}{8}$	$\dfrac{16}{81} + \dfrac{1}{81} = \dfrac{17}{81}$	$\dfrac{81}{256} + \dfrac{1}{256} = \dfrac{41}{128}$	$\dfrac{81}{65} + \dfrac{16}{625} = \dfrac{97}{625}$

(f) When the teams are evenly matched (as with 1:1 odds), the chances of the series ending in four games is small (0.125) compared to when the teams are not evenly matched.

Section 11.3

22. (a) Assuming independence of games, each of the four ways the American League can win in five games is a combination of four wins and one loss. Therefore, when calculating the probability for any of the four events, there would be a product of 4 p's (wins) and 1 q (loss). The probability of each of the events is p^4q, so the probability that the American League wins in five games is $4p^4q$.

    (b) The four ways for the National League to win the series in five games are ANNNN, NANNN, NNANN, and NNNAN. The probability of each event is found by multiplying 4 q's (for 4 National League wins) and $1p$ (for 1 National League loss). Thus, the probability of each event is q^4p, so the probability that the National League wins in five games is $4q^4p$.

    (c) Since the events are mutually exclusive, we add their probabilities. The probability the series will end after five games is $4p^4q + 4q^4p$.

Section 11.3

23. (a) Since there are ten ways the American League can win a six-game world series and each of these ways involves four wins and two losses, the probability calculation for each of the ten ways is the same. The probability of a win is P(A) = p, so P(N)= q. The probability that the American League wins a six-game series is $10p^4q^2$.

    (b) Similarly, the probability that the National League wins a six-game series is $10q^4p^2$.

    (c) Since these events are mutually exclusive, the probability that the series will end at six games is the sum of the probability that the American League wins in six games and the probability that the National League wins in six games: $10p^4q^2 + 10q^4p^2$.

Section 11.3

24. These probabilities follow the same pattern as in problems 22 and 23.

    (a) Probability that the American League wins in 7 is $20p^4q^3$.

    (b) Probability that the National League wins in 7 is $20q^4p^3$.

    (c) Probability that the World Series goes all 7 games is $20p^4q^3 + 20\,q^4p^3$.

Section 11.3

25. (a) Consider our work from problems 21 through 24.

x = number of games	4	5	6	7
P(American wins)	p^4	$4p^4q$	$10p^4q^2$	$20p^4q^3$
P(National wins)	q^4	$4q^4p$	$10q^4p^2$	$20q^4p^3$
P(x games in series)	$p^4 + q^4$	$4p^4q + 4q^4p$	$10p^4q^2 + 10q^4p^2$	$20p^4q^3 + 20q^4p^3$

(b) Since the odds favoring the American League are 1:1,
$p = \dfrac{1}{1+1} = \dfrac{1}{2}$, so $q = 1 - \dfrac{1}{2} = \dfrac{1}{2}$. Consider the row labeled P(x games in series).

x = Number of Games	P(x)
4	$\left(\dfrac{1}{2}\right)^4 + \left(\dfrac{1}{2}\right)^4 = \dfrac{1}{8}$
5	$4\left(\dfrac{1}{2}\right)^4\left(\dfrac{1}{2}\right) + 4\left(\dfrac{1}{2}\right)^4\left(\dfrac{1}{2}\right) = \dfrac{1}{4}$
6	$10\left(\dfrac{1}{2}\right)^4\left(\dfrac{1}{2}\right)^2 + 10\left(\dfrac{1}{2}\right)^4\left(\dfrac{1}{2}\right)^2 = \dfrac{5}{16}$
7	$20\left(\dfrac{1}{2}\right)^4\left(\dfrac{1}{2}\right)^3 + 20\left(\dfrac{1}{2}\right)^4\left(\dfrac{1}{2}\right)^3 = \dfrac{5}{16}$

(c) Let v_i = number of games and p_i = the probability that the series ends in v_i games. Then the expected value is

$$v_1p_1 + v_2p_2 + v_3p_3 + v_4p_4 = 4 \times \dfrac{1}{8} + 5 \times \dfrac{1}{4} + 6 \times$$

$$\dfrac{5}{16} + 7 \times \dfrac{5}{16} \approx 5.8 \text{ games.}$$

Section 11.3

26. We can simulate buying snack boxes until we have five different prizes by rolling a standard die until each of the numbers 1 through 5 appears and then counting the number of rolls needed. (Note: Ignore the 6 whenever it comes up. Do not count it). Getting all five different prizes would constitute a trial in the experiment. Repeat the experiment at least 100 times. We conducted our own simulation, with results as summarized in the following table.

Number of Rolls Needed to Get All 5 Prizes	Number of Trials out Of 100	Probability
5	3	0.03
6	4	0.04
7	5	0.05
8	8	0.08
9	10	0.10
10	10	0.10
11	16	0.16
12	12	0.12
13	8	0.08
14	6	0.06
15	3	0.03
16	4	0.04
17	2	0.02
18	1	0.01
19	2	0.02
20	2	0.02
21	0	0
22	1	0.01
23	1	0.01
24	0	0
25	0	0
26	1	0.01
27	1	0.01

Expected value = 5(0.03) + 4(0.04) + 5(0.05) + ... + 26(0.01) + 27(0.01) = 11.73. Your expected value should be approximately 11.5 since the theoretical expected value is 11.42.

SOLUTIONS - PART A PROBLEMS

Chapter 12: Geometric Shapes

Section 12.1

12. (a) To change from A to B, the diagonal of the square has been constructed between the lower left and the upper right vertices. The figure X is a rectangle, and it belongs with choice (ii) since (ii) is the only rectangle with a diagonal.

(b) To change from A to B, the rhombus was rotated one-half turn. The figure X is a square with a dot in the right-hand triangle formed by the two diagonals. X belongs with choice (i) since it is a square with a dot in the left-hand triangle, which resulted from a rotation of one-half turn.

(c) To change from A to B, the triangle was reflected over a vertical line. If figure X is reflected over a vertical line, the triangle ends up on the right-hand side of the square. Therefore, X belongs with choice (i).

(d) To change from A to B, the square was rotated clockwise 45° about its center. This moves the shaded triangle to the right-hand side of the square. Therefore, X belongs with choice (ii).

Section 12.1

13. (a) If you use a ruler or protractor to attempt to draw parallel lines on dot paper, then notice that it can be done by drawing lines along two rows of dots or along any two columns of dots. Notice that other parallel lines can also be drawn.

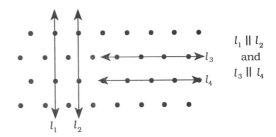

$$l_1 \parallel l_2$$
and
$$l_3 \parallel l_4$$

(b) If you attempt to draw perpendicular lines, you will see that by drawing a line along a row (horizontally) and a column (vertically), perpendicular lines result. Notice that other perpendicular lines are possible. In part (a), l_2 is perpendicular to l_3, l_1 is perpendicular to l_3, l_2 is perpendicular to l_4, and l_1 is perpendicular to l_4.

Section 12.1

14. (a) A rhombus is a quadrilateral having all four sides equal. Notice that in the triangular lattice, the distances from any dot to every adjacent dot are the same. Connecting four adjacent dots in the following manner forms a rhombus.

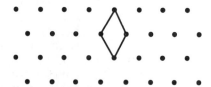

(b) Since every rhombus is also a parallelogram, and we can draw a rhombus on a triangular lattice, we can draw a parallelogram on a triangular lattice.

(c) Perpendicular lines can be drawn by constructing lines horizontally and vertically. Let x be the distance between adjacent dots. Consider any vertical column of dots. The distance between adjacent dots can be determined by the Pythagorean Theorem to be $x\sqrt{3}$. (See figure.)

$$x^2 + b^2 = (2x)^2$$
$$b^2 = 4x^2 - x^2$$
$$b^2 = 3x^2$$
$$b = x\sqrt{3}$$

There is no whole number of length x that equals a whole number multiple of $x\sqrt{3}$. Therefore, a square, which is a quadrilateral with four right angles and equal side lengths, cannot be drawn on the triangular lattice. Similar reasoning holds for perpendicular lines that are not horizontal and vertical.

(d) We can construct a rectangle on a triangular lattice since four equal side lengths are not required. Four right angles are required, so perpendicular lines are needed. The sides of the rectangle can be horizontal and vertical. Choose any two dots in any row and draw vertical lines through them. Choose any two dots along one of the vertical lines, and draw horizontal lines through them. The intersection points of the lines form the vertices of the rectangle. Notice that many other solutions are possible.

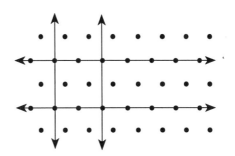

Section 12.1

15. (a) Consider the first two rows of dots. The following three parallelograms can be drawn:

    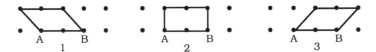

    Similar parallelograms can be drawn for each of the other three rows of dots. Therefore, there is a total of (4 rows) × (3 parallelograms) = 12 parallelograms.

    (b) Since rectangles are parallelograms that also have four right angles, none of the slanted parallelograms found in part (a) are rectangles. Therefore, only parallelograms that use dots directly above or below dots A and B form rectangles. There is only one rectangle formed with each of the other four rows. Therefore, four rectangles can be drawn.

    (c) A rhombus has four equal side lengths. Thus, neither row adjacent to side $\overline{AB}$ will produce a rhombus since those rows are less than the length AB (2 units) away.

    Also, row 5 is more than 2 units from side $\overline{AB}$. Only in row 4 will we find possible vertices for a rhombus. Connect points A and B to dots directly below them in row 4. Each side will be 2 units in length. This figure is a rhombus. Connecting A and B to any other dots in row 4 produces a slanted quadrilateral whose side length is $\sqrt{5}$ units. Therefore, no slanted figure will be a rhombus. Only one rhombus can be drawn on the lattice.

    (d) Every square is a rhombus, and we found that there can be at most one rhombus. Since the rhombus in part (c) has four sides of length 2 units and four right angles, it is a square. Therefore, there is one square.

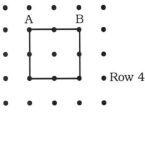

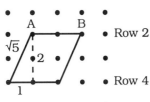

Section 12.1

16. (a) Consider the center of the hexagon. Draw a rhombus so that two adjacent sides $\overline{AB}$ and $\overline{AF}$ lie on the hexagon, and the opposite vertex is at the center of the hexagon. By connecting the center of the hexagon to vertices B and F, one rhombus is formed. If you construct the segment that connects the center of the hexagon to vertex D, two more rhombuses will be formed.

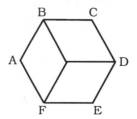

(b) Connect the center of the hexagon to the midpoints of each of the six sides. Six kites will be formed. Notice that the kites are not rhombuses.

Section 12.2

8. Systematically rearrange squares to form tetrominos.

Consider four squares in a row. There is only one such tetromino.

Consider three squares in a row and move the fourth to form two more tetrominos.

Consider two squares in a row and move the other two to form two more tetrominos.

There are a total of five tetrominos.

Section 12.2

9. Construct two copies of each tetromino on graph paper and cut out the pieces. Draw a 5 x 8 rectangle. The pieces fit together as shown in the next figure.

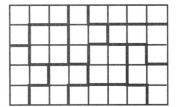

Section 12.2

10. (a) Recall the definitions of each type of triangle. An isosceles triangle has two or three sides the same length. A scalene triangle has three sides of different lengths. Since a scalene triangle cannot have any sides the same length, a scalene triangle is never an isosceles triangle. Since an isosceles triangle must have at least two sides the same length, an isosceles triangle is never a scalene triangle. The sets are thus disjoint. Option (i) represents the relationship.

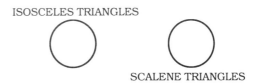

ISOSCELES TRIANGLES

SCALENE TRIANGLES

ISOSCELES TRIANGLES

EQUILATERAL TRIANGLES

(b) Since isosceles triangles have *at least* two sides of equal length, and equilateral triangles have three equal sides, every equilateral triangle is an isosceles triangle. Therefore, the set of equilateral triangles is a subset of the set of isosceles triangles. Option (iii) represents the relationship.

(c) In a right triangle, the two legs forming the 90° angle could be equal in length. In that case, the right triangle would be an isosceles triangle. Since equilateral triangles are isosceles triangles with three 60° angles, there are triangles that are isosceles but not right. Therefore, the sets intersect, but neither set is a subset of the other set. Option (ii) represents the relationship.

(d) Equilateral triangles always have three 60° angles. Since no equilateral triangle can have a 90° angle, the set of equilateral triangles cannot intersect the set of right triangles. Option (i) represents the relationship.

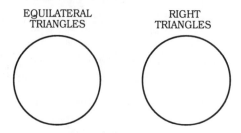

Section 12.2

11. Copy rectangle ABCD, including the diagonals. Flip the tracing and place it over the original so that the vertices of the original and the tracing correspond in the following way: A to D, B to C, C to B, and D to A. Notice that the diagonals are the same length.

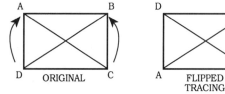

Section 12.2

12. Copy the square. When we perform the rotation, we will rotate the figure clockwise about its center. If we rotate square ABCD one-quarter turn, we will notice that the traced image coincides with the original square. Notice also that the tracing coincides with the original square after a $\frac{1}{2}, \frac{3}{4}$, and full turn. Therefore, the rotations of symmetry of a square are $\frac{1}{4}, \frac{1}{2}, \frac{3}{4}$, and 1 full turn.

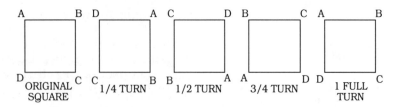

Section 12.2

13. Copy the equilateral triangle. The center of rotation is the intersection of the reflection lines. Label the triangle and rotate it clockwise $\frac{1}{3}$ of a turn until the traced triangle coincides with the original triangle. A $\frac{2}{3}$ turn and a full turn also cause the tracing to coincide with the original triangle. Therefore, the rotations of symmetry of an equilateral triangle are $\frac{1}{3}, \frac{2}{3}$, and 1 full turn.

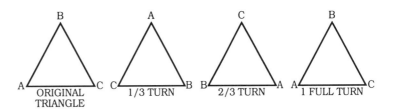

Section 12.2

14. (a) Fold kite ABCD along the diagonal so that vertices D and B coincide. Notice that ∠ADC coincides with ∠ABC. To determine if the other angles are congruent, fold the original kite so that the vertices A and C coincide. When this has been done, ∠DAB is larger than ∠DCB. Therefore, these angles are not congruent.

    (b) A kite is a quadrilateral with two, nonoverlapping pairs of adjacent sides congruent. We saw from part (a) that two of the opposite angles of kite are congruent. A rhombus has all four sides congruent. Therefore, it has four pairs of adjacent sides congruent. A rhombus is a kite in two ways, since four pairs of adjacent sides are congruent rather than just two. Therefore, both pairs of opposite angles are congruent.

Section 12.2

15. Make a tracing of the given rhombus. Take your tracing and rotate it $\frac{1}{2}$ turn around the center E. Then $\overline{AE}$ in your tracing should coincide with $\overline{CE}$ from the original diagram. Also, $\overline{BE}$ in your tracing coincides with $\overline{DE}$ from the original diagram. Therefore, we have that $\overline{AE} \cong \overline{CE}$ and $\overline{BE} \cong \overline{DE}$, so the diagonals of a rhombus bisect each other.

Section 12.2

16. Consider rhombus ABCD. Label the intersection of the diagonals as point E. To show that $\overline{AC}$ is perpendicular to $\overline{DB}$, fold the rhombus along diagonal $\overline{DB}$, as shown in the figure, and notice that $\overline{AE}$ and $\overline{CE}$ coincide. Therefore, $\overline{AC} \perp \overline{DB}$.

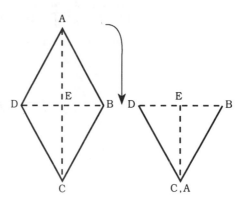

Section 12.2

17. (a) An isosceles triangle must have at least two legs of equal length. Consider the possible number of toothpicks for each leg. Realize that for a triangle to exist, the sum of the lengths of any two sides must be greater than the length of the third side. Consider the following possibilities for 20 toothpicks.

Leg	Leg	Base	
1	1	18	Since 18 > 1 + 1, impossible
2	2	16	Since 16 > 2 + 2, impossible
3	3	14	Since 14 > 3 + 3, impossible
4	4	12	Since 12 > 4 + 4, impossible
5	5	10	Since 10 = 5 + 5, impossible
6	6	8	Possible
7	7	6	Possible
8	8	4	Possible
9	9	2	Possible
10	10	0	Impossible

Therefore, the only possible isosceles triangles that can be formed using 20 toothpicks have legs of lengths 6-6-8, 7-7-6, 8-8-4, or 9-9-2. Consider the possibilities for 24 toothpicks.

Leg	Leg	Leg	
1	1	22	
2	2	20	
3	3	18	
4	4	16	
5	5	14	
6	6	12	
7	7	10	Possible
8	8	8	Possible
9	9	6	Possible
10	10	4	Possible
11	11	2	Possible
12	12	0	

Therefore, the only possible isosceles triangles that can be formed using 24 toothpicks have legs of lengths 7-7-10, 8-8-8, 9-9-6, 10-10-4, or 11-11-2.

(b) Scalene triangles must have three different side lengths. Since no one side can be greater than the sum of the lengths of the other two sides, no side can be greater than or equal to half the number of toothpicks. For example, using 20 toothpicks, no side can be made of more than 9 toothpicks.

Longest Leg	Second Longest Leg	Shortest Leg	
9	8	3	
9	7	4	
9	6	5	
8	7	5	
8	6	6	(not scalene)

Therefore, the only scalene triangles that can be formed using 20 toothpicks have side lengths of 9-8-3, 9-7-4, 9-6-5, or 8-7-5. Using 24 toothpicks no side can be made of more than 11 toothpicks.

Longest Side	Second Longest Side	Shortest Side	
11	10	3	
11	9	4	
11	8	5	
11	7	6	
10	9	5	
10	8	6	
10	7	7	(not scalene)
9	8	7	

There are only seven different scalene triangles that can be formed using 24 toothpicks.

(c) Equilateral triangles have side lengths that are all the same. Each side length is $\frac{1}{3}$ of the number of toothpicks. Since $\frac{20}{3} = 6\frac{2}{3}$, no equilateral triangle can be formed using 20 toothpicks. Since $\frac{24}{3} = 8$, there is exactly one equilateral triangle that can be formed using 24 toothpicks. Each side must be made up of 8 toothpicks.

Section 12.3

8. Make a list of all known angles. We are given that $m\angle BFC = 55°$ and $m\angle AFD = 150° = m\angle AFB + m\angle BFC + m\angle CFD$. By substitution, we know
$$150° = m\angle AFB + 55° + m\angle CFD,$$
$$\text{so, } 95° = m\angle AFB + m\angle CFD.$$
We are also given that $m\angle BFE = 120°$.
Thus, $m\angle BFE = 120° = m\angle BFC + m\angle CFD + m\angle DFE$.
By substitution, we know $120° = 55° + m\angle CFD + (180° - 150°)$ since $\angle AFD$ and $\angle DFE$ are supplementary.
Therefore, $35° = m\angle CFD$ and $m\angle AFB = 95° - 35° = 60°$.

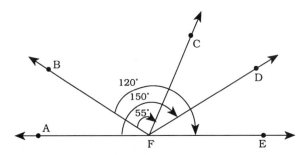

Section 12.3

9. Since the measure of $\angle 1$ is 9° less than half the measure of $\angle 2$, $m\angle 1 = \frac{1}{2}m\angle 2 - 9°$. Since $\angle 1$ and $\angle 2$ are supplementary, we know that $m\angle 1 + m\angle 2 = 180°$. By substitution, we know
$$\frac{1}{2}m\angle 2 - 9° + m\angle 2 = 180°.$$
$$\frac{3}{2}m\angle 2 = 189°$$
$$m\angle 2 = 126°$$
$$m\angle 1 = 180° - 126° = 54°$$
Therefore, $m\angle 1 = 54°$ and $m\angle 2 = 126°$.

Section 12.3

10. We are given that $m \angle 1 = 80°$. Then we know the following.

By vertical angles, $m \angle 11 = 80°$.
By supplementary angles, $m \angle 2 = 180° - 80° = 100°$.
By supplementary angles $m \angle 12 = 180° - 80° = 100°$.

We are given that $m \angle 4 = 125°$. Then we know the following.

By vertical angles, $m \angle 6 = 125°$.
By supplementary angles, $m \angle 5 = 180° - 125° = 55°$.
By supplementary angles, $m \angle 7 = 180° - 125° = 55°$.

Since m and l are parallel, $\angle 1$ and $\angle 16$ are corresponding angles. Then we know the following.

By corresponding angles, $m \angle 16 = 80°$.
By vertical angles, $m \angle 18 = 80°$.
By supplementary angles, $m \angle 17 = 180° - 80° = 100°$.
By supplementary angles, $m \angle 15 = 180° - 80° = 100°$.

Since m and l are parallel, $\angle 4$ and $\angle 13$ are corresponding angles. Then we also know the following.

By corresponding angles, $m \angle 13 = 125°$.
By vertical angles, $m \angle 19 = 125°$.
By supplementary angles, $m \angle 14 = 180° - 125° = 55°$.
By supplementary angles, $m \angle 20 = 180° - 125° = 55°$.

Since the measures of the angles in any triangle must add up to $180°$, $m \angle 10 + m \angle 11 + m \angle 14 = 180°$, so $m \angle 10 = 180° - 80° - 55° = 45°$ by substitution. Then we know the following.

By vertical angles, $m \angle 8 = 45°$.
By supplementary angles, $m \angle 9 = 180° - 45° = 135°$.
By supplementary angles, $m \angle 3 = 180° - 45° = 135°$.

Section 12.3

11. We are given that $m \angle 1 = m \angle 2$. By vertical angles, we know that $m \angle 2 = m \angle 3$. By substitution, we know $m \angle 1 = m \angle 3$, and thus $\angle 1 \cong \angle 3$. Notice that $\angle 1$ and $\angle 3$ are congruent corresponding angles. Therefore, $l \parallel m$, because corresponding angles are congruent.

Section 12.3

12. (a) Suppose $m \angle 1 = m \angle 6$. We know $m \angle 4 = m \angle 6$ by vertical angles. By substitution, we know $m \angle 1 = m \angle 4$, so $\angle 1 \cong \angle 4$. Notice that $\angle 1$ and $\angle 4$ are corresponding angles. Since corresponding angles are congruent, $l \parallel m$.

(b) If $l \parallel m$, then $m \angle 3 = m \angle 6$ by corresponding angles. By vertical angles, $m \angle 1 = m \angle 3$. By substitution, we have that $m \angle 1 = m \angle 6$.

(c) Two lines are parallel if, and only if, at least one pair of alternate exterior angles formed by the intersection of the two lines and the transversal have the same measure.

Section 12.3

13. Since $m \angle A$ is 4 times as large as $m \angle B$, we know $m \angle A = 4m \angle B$. Since $\angle A$ and $\angle B$ are complementary angles, we know that $m \angle A + m \angle B = 90°$.

Then, by substitution, $4m \angle B + m \angle B = 90°$
$$5m \angle B = 90°$$
$$m \angle B = \frac{90°}{5}$$
$$m \angle B = 18°$$
$$m \angle A = 4(18°) = 72°.$$

Section 12.3

14. Notice in the figure that $\angle 2$ and $\angle 3$ are corresponding angles. We know $m \angle 2 = 90°$ by supplementary angles, and $m \angle 3 = 90°$ by vertical angles. Since $m \angle 2 = 90° = m \angle 3$, we see that corresponding angles are congruent. Therefore, by the Corresponding Angles Property, lines $\overleftrightarrow{AB}$ and $\overleftrightarrow{DC}$ are parallel. Notice also that $\angle 1$ and $\angle 3$ are corresponding angles. $m \angle 3 = 90°$ by vertical angles, and $m \angle 1 = 90°$ by supplementary angles. Since $m \angle 1 = 90° = m \angle 3$, we see that corresponding angles are congruent. Therefore, by the Corresponding Angles Property, lines $\overleftrightarrow{AD}$ and $\overleftrightarrow{BC}$ are parallel. Since there are two pairs of opposite sides parallel, the rectangle ABCD is a parallelogram.

Section 12.3

15. (a) (i)   The figures do not intersect.

(i)

(ii) The figures intersect in one point.

(iii) The figures must overlap for them to intersect in two points.

(ii)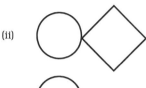

(iv) The figures must overlap as in part (iii). Continue moving the square into the circle until the vertex of the square just touches the inside of the circle. In this way, the figures will intersect in three points.

(b) We have drawn the figures so that they intersect in one, two, or three points. To draw figures that intersect in four points, inscribe the circle in the square or circumscribe the circle about the square.

(iii)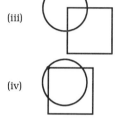

(iv)

CIRCLE CIRCUMSCRIBED

CIRCLE INSCRIBED

4 INTERSECTION POINTS

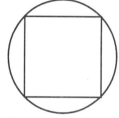

4 INTERSECTION POINTS

Consider the figure with the circle circumscribed about the square. If we reduce the radius of the circle, keeping the centers of both figures the same, eventually the circle will intersect each side of the square twice. This gives eight intersection points, two on each side of the square. Since there is no way for a circle to intersect the side of a square in more than two points, the maximum number of intersection points of a square and a circle is eight.

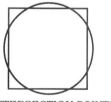

8 INTERSECTION POINTS

Section 12.3

16. (a) Refer to the figures that follow (i) - (iv).

(i)   Two disjoint figures have no intersection points.

(ii) One intersection point is shown.

(iii) The figures must overlap in order for there to be two points of intersection.

(iv) Consider the same configuration as in (iii). By moving the triangle into the circle until the vertex intersects the circle, we can draw figures that intersect in three points.

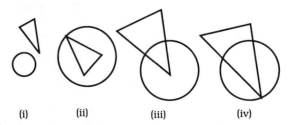

(i)   (ii)   (iii)   (iv)

(b) Consider other ways of obtaining three intersection points. One way is to circumscribe a circle about an equilateral triangle, as shown in the figure.

3 INTERSECTION POINTS

If the radius of the circle is reduced, eventually the circle will intersect each side of the triangle twice. Since a circle cannot intersect a line segment in more than two points, the maximum number of points of intersection for a circle and a triangle is six.

6 INTERSECTION POINTS

Section 12.3

17. Draw pictures and count the number of regions into which the plane is divided by each set of lines. Make a table. Keep in mind that two lines can intersect in at most one point. To maximize the number of regions, each new line should intersect each existing line once. When

introducing a third line, be sure to construct it so that it intersects each of the existing lines at different points.

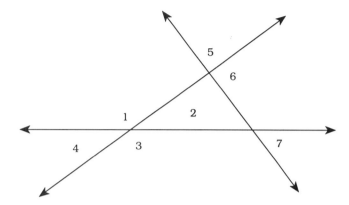

We see that three lines divide the plane into at most seven regions. Consider four lines. To maximize the number of regions, construct the fourth line so that it intersects each of the three existing lines in different places. Avoid previous intersection points.

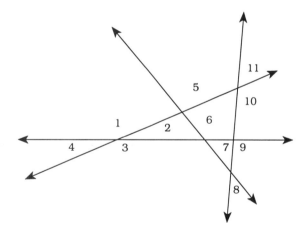

We see that four lines divide the plane into at most eleven regions. Now consider five lines. Construct the fifth line so that it intersects each of the four existing lines in different places. Avoid all previous intersection points.

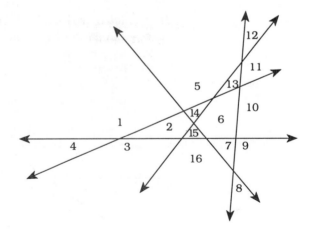

We see that five lines divide the plane into at most sixteen regions. We will summarize our results and look for a pattern in the table shown next.

Number of Lines	Number of Regions
1	$2 = 1 + 1$
2	$4 = 1 + 1 + 2$
3	$7 = 1 + 1 + 2 + 3$
4	$11 = 1 + 1 + 2 + 3 + 4$
5	$16 = 1 + 1 + 2 + 3 + 4 + 5$
.	
.	
.	
10	$56 = 1 + 1 + 2 + 3 + 4 + 5 + 6 + 7 + 8 + 9 + 10$
.	
.	
.	
n	$1 + 1 + 2 + 3 + 4 + 5 + ... + (n - 1) + n$ $= 1 + n\text{th triangular number}$ $= 1 + \dfrac{n(n + 1)}{2}$

Therefore, we see that the greatest number of regions into which the plane can be divided by n lines is $1 + \dfrac{n(n + 1)}{2}$.

Section 12.4

16. Consider a simpler problem. Suppose there were only four people in the room. Let the four people be represented by A, B, C, and D. Make a systematic list of all possible handshakes. Keep in mind that a handshake between A and B, "AB", is the same as a handshake between B and A, "BA". Avoid counting handshakes twice.

> Handshakes: AB
> AC   BC
> AD   BD   CD

When there are four people, there are 6 possible handshakes.

Suppose there are 5 people: A, B, C, D, and E.
> Handshakes: AB
> AC   BC
> AD   BD   CD
> AE   BE   CE   DE

When there are 5 people, there are 10 possible handshakes.

Number of People	Number of Handshakes
4	$6 = 1 + 2 + 3$
5	$10 = 1 + 2 + 3 + 4$

Notice a pattern in the number of handshakes. The number of handshakes possible among n people is the sum of the first $n - 1$ counting numbers. Therefore, if there are 20 people, the number of handshakes possible is

$$1 + 2 + 3 + \ldots + 19 = \frac{19 \times 20}{2} = 190 \text{ handshakes.}$$

Section 12.4

17. $\angle d$: Since the sum of the measures of the angles in a triangle is $180°$, $m\angle d = 180° - 110° - 50° = 20°$.

    $\angle e$: Since $\angle d$ and $\angle e$ make up one angle in a right triangle, and we know $m\angle d$, we know that $m\angle e = 180° - 50° - 90° - m\angle d = 180° - 50° - 90° - 20° = 20°$.

    $\angle c$: Since $\angle c$ is supplementary with a $60°$ angle, we know that $m\angle c = 180° - 60° = 120°$.

    $\angle g$: Since $\angle g$ is congruent, by vertical angles, to $\angle x$ and we know $m\angle x = 180° - 90° - 30° = 60°$, $m\angle g = 60°$.

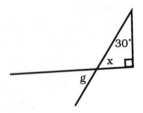

$\angle f$: $m\angle f = 180° - 40° - 60° = 80°$.

$\angle a$: Since $\angle a$ is congruent to $\angle y$, we know $m\angle a = m\angle y$.

$$m\angle a + m\angle y + 40° = 180°$$
$$2m\angle a = 140°$$
$$m\angle a = 70°$$

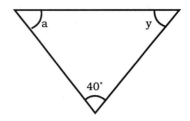

$\angle b$: Since $\angle b$ is supplementary to $\angle z$ and $m\angle z = 180° - 60° - m\angle a$, we know $m\angle z = 180° - 60° - 70° = 50°$. Therefore, $m\angle b = 180° - 50° = 130°$.

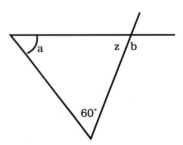

$\angle h$: Since we know $m\angle b$, we know the measure of $\angle w$ by supplementary angles. We know that $m\angle w = 180° - m\angle b = 180° - 130° = 50°$ and $m\angle t = 180° - 50° - 50° = 80°$. Since $\angle h$ is supplementary to $\angle t$, we know that $m\angle h = 180° - 80° = 100°$.

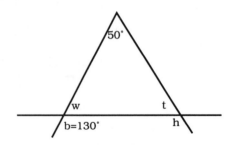

Section 12.4

18. The center figure in the star is a regular pentagon. Each vertex angle measures 108° by the Angle Measure Theorem in a regular *n*-gon. By supplementary angles, each base angle in each triangle has measure 180° − 108° = 72°. Since the sum of the angles in any triangle is 180°, each angle of the star (∠A, ∠B, ∠C, ∠D, and ∠E) measures 180° − 72° − 72°= 36°. Therefore, the sum of the angle measures in the star is *m*∠A + *m*∠B + *m*∠C + *m*∠D + *m*∠E = 5(36°) = 180°.

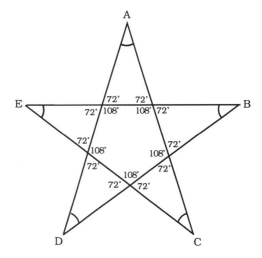

Section 12.5

24. Make a model and compare each of the choices given to your model. When comparing your cube to the choices, be sure your model has the same orientation. That is, be sure you can see the front, top, and left side.
    (a) The net will fold to become cube (iii).
    (b) The net will fold to become cube (ii).

Section 12.5

25. (a) Since there are three sets of opposite faces, there are three planes of symmetry of this type.

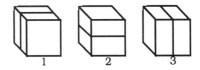

(b) There are twelve edges on a cube, which means there are six sets of opposite edges. Therefore, there are six planes of symmetry of this type.

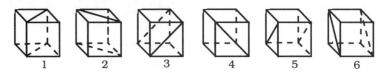

(c) The only planes of symmetry result from cutting the cube midway between opposite faces or from cutting through pairs of opposite edges. Since there are three and six of these types of symmetries, respectively, there are only nine planes of symmetry for a cube.

Section 12.5
26. The axis of rotation passes through pairs of opposite faces of the cube. Since there are three pairs of opposite faces, there are three axes of symmetry of order four in a cube.

Section 12.5
27. (a) Three faces of the cube shown in the text meet at one vertex. When the cube is in its original position, one of these faces is showing. (Face E is the one showing in the picture when the cube is in its original position.) As the cube is rotated on the axis, the "original" arrangement appears each time another face rotates into position E. Therefore, since there are three faces that can occupy position E, the order is three.
(b) Since there are eight vertices, there are four sets of opposite pairs. Therefore, there are four axes of symmetry for this cube of order three.

Section 12.5
28. (a) The axis of rotation passes through the midpoints of opposite edges. For any single edge, two faces meet at that edge. When the cube is in its original position, one of these two faces is showing. As the cube is rotated on the axis, it will only return to its original position when the other of the two faces is showing. Therefore, since there are two faces under consideration, the order is two.
(b) Since there are twelve edges on a cube, there are six pairs of opposite edges. Therefore, there are six possible axes of symmetry of this type.

Section 12.5

29. The center tube for a paper towel roll is a cylinder. Notice the spiral seam lines on any paper towel roll. Cut along the seam line. The piece of cardboard from which the tube was made is a long parallelogram.

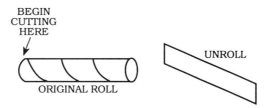

Section 12.5

30. (a) Yes. Since the box pictured is a rectangular prism, we could stack copies of it on top of one another with no gaps. We could also lay them end to end and side to side with no gaps. Therefore, we can fill up space with no gaps or overlap.

(b) Recall the discussion about tessellations from Section 12.4. We know that any triangle, quadrilateral, or regular hexagon will tessellate the plane. Each of these plane figures can be made three-dimensional by forming a right prism with the polygon as a base. Copies of any one of the prisms could then be stacked to fill space.

(c) Recall that the ancient Greeks showed that there are exactly five regular convex polyhedra. In order for a three-dimensional figure to tessellate space, it must be a prism. Since the only regular polyhedron that is also a prism is a hexahedron (cube), there is only one regular polyhedron that will tessellate space by itself.

SOLUTIONS - PART A PROBLEMS

Chapter 13: Measurement

Section 13.1

26. Use dimensional analysis and the fact that 1 gallon of water weighs about 8.3 pounds to convert 1 cubic foot into gallons. We know that 1 cubic foot weights 62 pounds, so we can convert 62 pounds into gallons.

$$62 \text{ pounds } \times \frac{1 \text{ gallon}}{8.3 \text{ pounds}} = \frac{62}{8.3} \text{ gallons } \approx 7.5 \text{ gallons.}$$

Therefore, there are about 7.5 gallons in one cubic foot of water.

Section 13.1

27. To find the approximate number of red blood cells in the body of an adult male, use the fact that there are about 5.4×10^6 cells per microliter (μL) of blood. The 70-kg man has a blood volume of about 5 liters. Recall that there are 1 million microliters in 1 liter.

$$5 \text{ L} \times \frac{1,000,000 \, \mu\text{L}}{1 \text{ L}} \times \frac{5.4 \times 10^6 \text{ red blood cells}}{1 \, \mu\text{L}} = 2.7 \times 10^{13}.$$

Therefore, there are about 2.7×10^{13} red blood cells in the body of an adult male.

Section 13.1

28. (a) Recall that there are 1000 grams in a kilogram, and there are 10 centimeters in 1 decimeter, so there are 1000 cm^3 in 1 dm^3.

$$\frac{8.94 \text{ g}}{1 \text{ cm}^3} \times \frac{1000 \text{ cm}^3}{1 \text{ dm}^3} \times \frac{1 \text{ kg}}{1000 \text{ g}} = 8.94 \frac{\text{kg}}{\text{dm}^3}.$$

The density of copper is 8.94 kg/dm^3.

(b) Since density = $\dfrac{\text{mass}}{\text{volume}}$, the wood has a density of $\dfrac{2.85 \text{ kg}}{4100 \text{ cm}^3}$. Convert to grams/cm^3 using dimensional analysis.

$$\frac{2.85 \text{ kg}}{4100 \text{ cm}^3} \times \frac{1000 \text{ g}}{1 \text{ kg}} \approx 0.695 \frac{\text{g}}{\text{cm}^3}.$$

The density of oak is about 0.695 g/cm^3.

(c) Recall that 16 ounces = 1 pound, 2.2 pounds = 1 kilogram, 1 kilogram = 1000 grams, and 2.54 centimeters = 1 inch. So 16.387 $cm^3 \approx 1$ $inch^3$.

$$\text{Density} = \frac{45 \text{ ounces}}{10 \text{ inches}^3}$$

$$\frac{45 \text{ oz}}{10 \text{ in}^3} \times \frac{1 \text{ lb}}{160 \text{ oz}} \times \frac{1 \text{ kg}}{2.2 \text{ lb}} \times \frac{1000 \text{ g}}{1 \text{ kg}} \times \frac{1 \text{ in}^3}{16.387 \text{ cm}^3} \approx 7.8 \frac{g}{cm^3}$$

The density of iron is about 7.8 g/cm^3.

Section 13.1

29. The English system is not portable since it cannot be reproduced without reference to a prototype. The natural English units were standardized so that the foot was defined by a prototype metal bar. Because of this, the foot cannot be reproduced anywhere there is no prototype available. In the English system, there are not simple (decimal) ratios among units of the same type. This means it is not easily convertible. The natural progression of units in the English system is inches, feet, yards, and miles. There are 12 inches in 1 foot, 3 feet in 1 yard, and 1760 yards in 1 mile.

There is no direct interrelatedness in the English system. Conversions would be relatively simple if cubic inches, quarts, and pounds were related because they are basic units of length, volume, and weight.

Section 13.1

30. (a) We would like to convert miles per second to miles per year using conversion ratios.

$$\frac{186282 \text{ miles}}{1 \text{ second}} \times \frac{60 \text{ seconds}}{1 \text{ minute}} \times \frac{60 \text{ minutes}}{1 \text{ hour}} \times \frac{24 \text{ hours}}{1 \text{ day}} \times \frac{365 \text{ days}}{1 \text{ year}} \approx$$
$$5.8746 \times 10^{12} \frac{\text{miles}}{\text{year}}$$

Therefore, the distance that light travels in one year is approximately 5.8746×10^{12} miles.

(b) Since 1 light year $\approx 5.8746 \times 10^{12}$ miles, and the star in Andromeda is 76 light years away, the light will travel

$$76 \text{ years} \times \frac{5.8746 \times 10^{12} \text{ miles}}{1 \text{ light year}} \approx 4.4647 \times 10^{14} \text{ miles}.$$

(c) Since we know 1 light year $\approx 5.8746 \times 10^{12}$ miles, we will use this conversion ratio to convert 480,000,000 miles to years.

$$480{,}000{,}000 \text{ miles} \times \frac{1 \text{ year}}{5.8746 \times 10^{12} \text{ miles}} \approx$$

$$0.000081708 \text{ years.}$$

$$0.000081708 \text{ years} \times \frac{365 \text{ days}}{1 \text{ year}} \approx$$

$$0.029823362 \text{ days} \times \frac{24 \text{ hours}}{1 \text{ day}} \approx$$

$$0.71576 \text{ hours} \times \frac{60 \text{ minutes}}{1 \text{ hour}} \approx 42.9456 \text{ minutes.}$$

Therefore, it takes approximately 43 minutes for the light to travel from the sun to Jupiter.

Section 13.1

31. If you are a passenger in a train moving 50 mph, and you observe a train traveling in the opposite direction at 50 mph, it will seem as though the train is passing you by at 100 mph. We need to determine how far a train traveling at 100 mph would go in 5 seconds. If we convert the speed to feet per second, we can determine how far the train travels in feet.

$$\frac{100 \text{ miles}}{1 \text{ hour}} \times \frac{5280 \text{ feet}}{1 \text{ mile}} \times \frac{1 \text{ hour}}{3600 \text{ seconds}} = 146\frac{2}{3}\frac{\text{feet}}{\text{second}}.$$

Since rate × time = distance, we know that the length of the train is $146\frac{2}{3}\dfrac{\text{feet}}{\text{second}}$ × 5 seconds = $733\frac{1}{3}$ feet.

Section 13.1

32. (a) To determine the number of cubic inches of water, we need to convert 1 acre of ground into square inches.

$$1 \text{ acre} = 43{,}560 \text{ ft}^2 \times \frac{144 \text{ in}^2}{1 \text{ ft}^2} = 6{,}272{,}640 \text{ in}^2.$$

Since 1 inch of rain fell, the number of cubic inches is $6{,}272{,}640 \text{ in}^2 \times 1 \text{ in.} = 6{,}272{,}640 \text{ in}^3$. To determine what this volume is in cubic feet, use the fact that 1 acre = 43,560 ft^2. We know that 1 inch = $\dfrac{1}{12}$ foot, so

1 inch of rain covering 1 acre = $43{,}560 \text{ ft}^2 \times \dfrac{1}{12} \text{ ft} = $ 3630 ft^3.

(b) One cubic foot of water weighs approximately 62 pounds. Use this fact to convert cubic feet to pounds. From part (a), we know that one inch of rain covering one acre of ground has a volume of 3630 ft^3.

Therefore, the water weight is

$$3630 \text{ ft}^3 \times \frac{62 \text{ pounds}}{1 \text{ ft}^3} = 225{,}060 \text{ pounds.}$$

(c) One gallon of water weighs approximately 8.3 pounds. From part (b), we know that 225,060 pounds of water fell. Therefore, $225{,}060 \text{ pounds} \times \frac{1 \text{ gallon}}{8.3 \text{ pounds}} \approx 27{,}116$ gallons of water fell.

Section 13.1

33. (a) If the ruler had marks at only 1, 4, and 6, then it would look like the ruler below. (Note: 8 is automatically marked.)

To measure a length of
1, use the distance from the left end to 1.
2, use the distance from 4 to 6.
3, use the distance from 1 to 4.
4, use the distance from the left end to 4.
5, use the distance from 1 to 6.
6. Use the distance from the left end to 6.
7, use the distance from 1 to 8.
8, use the distance from the left end to 8.

(b) To measure a length of 1, we could have a mark one unit from either end. Place a mark one unit from the left end. Notice that we can measure a length of 1, a length of 8 (by using the distance from 1 to 9) and a length of 9 (by using the whole ruler). To measure a length of 2, place a mark three units from the left. Notice that we can measure a length of 2 (by using the distance from 1 to 3), a length of 3 (by using the distance from the left end to 3), and a length of 6 (by using the distance from 3 to 9). Placing a mark five units from the left allows us to measure a length of 4 (by using 1 to 5) and a length of 5 (by using the left end to 5). Finally, to measure a length of 7, place a mark seven units from the left.

Our solution is one of several possible solutions which use the minimum of 4 marks. There are others, such as placing marks at 1, 2, 3, and 5; placing marks at 1, 3, 5, and 8; or placing marks at 1, 4, 6, and 7.

(c) On a 10-unit ruler, place a mark 1 unit from the left. This will allow us to measure lengths of 1, 9, and 10. If a mark is placed 3 units from the left, then we will be able to measure lengths of 2, 3, and 7. If we place a mark 6 units from the left, then we will be able to measure lengths of 4, 5, and 6. A mark placed 8 units from the left will allow us to measure a length of 8.

Another solution using the minimum of four marks has marks at 1, 4, 6, and 8.

Section 13.1

34. (a) Together they cut 48 ft³ in 1 hour. There are 4 × 4 × 8 = 128 ft³ in a cord of wood. Using conversion ratios, we can convert from 48 ft³ per hour to dollars per day.

$$\frac{48 \text{ ft}^3}{1 \text{ hour}} \times \frac{8 \text{ hours}}{1 \text{ day}} \times \frac{1 \text{ cord}}{128 \text{ ft}^3} \times \frac{\$100}{1 \text{ cord}} = \$300 \text{ per day.}$$

(b) If they split the money evenly, the son would earn $150 for 8 hours of work or $\frac{\$150}{8 \text{ hours}} = \$18.75$ per hour.

(c) In one day they cut $\frac{48 \text{ ft}^3}{1 \text{ hour}} \times \frac{8 \text{ hours}}{1 \text{ day}} = \frac{384 \text{ ft}^3}{1 \text{ day}}$.
Since the truck can hold 100 ft³ per trip, it would take 4 trips to deliver all the wood cut in a day. Notice that 3 truck loads are full, and the last truck carries only 84 ft³.

(d) Since the truck holds 100 ft³ of wood, and they sell it for $85, we can convert from dollars per cubic foot to dollars per cord.

$$\frac{\$85}{100 \text{ ft}^3} \times \frac{128 \text{ ft}^3}{1 \text{ cord}} = \$108.80 \text{ per cord.}$$

Section 13.1

35. Let d = the distance from the bottom of the hill to the summit. Since distance = rate × time, we know that $T = \frac{D}{R}$. The time uphill $= \frac{d}{2}$. The time downhill $= \frac{d}{6}$. The total time for the entire trip is $\frac{d}{2} + \frac{d}{6} = \frac{2d}{3}$. The total distance for the entire trip is $2d$. Therefore, the rate for the entire trip is given by

$$\frac{\text{distance for entire trip}}{\text{time for entire trip}} = \frac{2d}{2d \, / \, 3} = 3 \, \frac{\text{km}}{\text{hour}} \cdot$$

Thus, the average speed for the trip is 3 km/hr.

Section 13.2

25. Notice that the wall of the building, the ground, and the ladder form a right triangle, since the ground and the wall of the building are perpendicular. Also, recall that any multiple of a 3-4-5 right triangle is a right triangle. Since $15 = 5 \times 3$ and $20 = 5 \times 4$, we know that $x = 5 \times 5 = 25$ feet. Therefore, the ladder is 25 feet long.

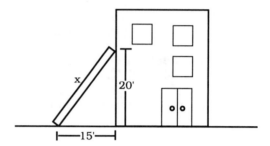

Section 13.2

26. Recall that the area of a rectangle is length × width. Area $= x \times y$. Since 96 meters of fencing are used, we know that $2x + y = 96$. Make a table of whole-number values of x and y that satisfy $2x + y = 96$ and compare areas.

x	y	$2x + y$	$A = x \times y$
1	94	96	94
2	92	96	184
3	90	96	270
4	88	96	352
5	86	96	430
6	84	96	504
7	82	96	574
8	80	96	640
9	78	96	702
10	76	96	760
11	74	96	814
12	72	96	864
13	70	96	910
14	68	96	952
15	66	96	990
16	64	96	1024
17	62	96	1054
18	60	96	1080
19	58	96	1102
20	56	96	1120
21	54	96	1134
22	52	96	1144
23	50	96	1150
24	48	96	1152
25	46	96	1150

Notice that areas increase until the width is exactly half as big as the length. After that point, the areas decrease. Therefore, the whole-number dimensions that yield the largest area are $x = 24$ m and $y = 48$ m.

Section 13.2

27. Consider a picture, as shown. Since the plane has a wingspan of 7.1 feet, it will have to be carried diagonally through the door, which is only $6\frac{1}{2}$ feet tall. Let the diagonal of the door have length c. Notice that the diagonal is the hypotenuse of a right triangle with legs 3 feet and $6\frac{1}{2}$ feet. By the Pythagorean Theorem $3^2 + (6.5)^2 = c^2$, so $9 + 42.25 = 51.25 = c^2$. Therefore, $c \approx 7.159$ feet. He should just be able to get the plane through the door.

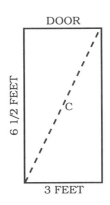

DOOR

6 1/2 FEET

c

3 FEET

Section 13.2

28. The area of the large figure is the area of a square with side length c. Area $= c^2$. Consider one of the right triangles. The legs of the triangle are a and b, so the area is $\frac{1}{2}ba$. Four right triangles have an area of $4\left(\frac{1}{2}ab\right) = 2ba$. The small square has side lengths $b - a$, so its area is $(b - a)^2$. The sum of the areas of the four triangles and the small square is equal to the area of the large square figure. So we have

$$c^2 = 4\left(\frac{1}{2}ab\right) + (b - a)^2$$
$$c^2 = 2ab + b^2 - 2ab + a^2$$
$$c^2 = a^2 + b^2.$$

BEHOLD!

Section 13.2

29. In the diagram, let m be the length of $\overline{AB}$ and n be the length of $\overline{AC}$.

(a) By the Pythagorean Theorem, we know $l^2 + w^2 = m^2$. Therefore, $m = \sqrt{l^2 + w^2}$.

(b) Notice that $\triangle ABC$ is a right triangle with sides m and h and hypotenuse n. By the Pythagorean Theorem, $m^2 + h^2 = n^2$. Therefore, $n = \sqrt{m^2 + h^2}$. From part (a) since $l^2 + w^2 = m^2$, $n = \sqrt{l^2 + w^2 + h^2}$.

(c) A rectangular box that has a width (w) of 40 cm, length (l) of 60 cm, and height (h) of 20 cm, has the length of the longest diagonal given by $n = \sqrt{l^2 + w^2 + h^2}$ from part (b). Therefore, the longest diagonal has a length of $\sqrt{60^2 + 40^2 + 20^2} = \sqrt{5600} \approx$ 74.8 cm.

Section 13.2

30. An old trunk can be considered a rectangular prism with length 30 inches, width 16 inches, and height 12 inches. From problem 29 we know the length of the longest diagonal is $\sqrt{l^2 + w^2 + h^2} = \sqrt{30^2 + 16^2 + 12^2} = \sqrt{1300} \approx$ 36.06 cm.

(a) A telescope measuring 40 inches would *not* fit diagonally into the trunk.

(b) A baseball bat measuring 34 inches would fit diagonally into the trunk.

(c) A tennis racket measuring 32 inches would fit diagonally into the trunk.

Therefore, Jason could store the baseball bat and the tennis racket in the trunk.

Section 13.2

31. In the right triangle given, b is the base, and a is the height. The formula for the area of this right triangle is:

$$A = \frac{1}{2}ba.$$

(a) Since A = 413.34 and a = 24.9 we can use the area formula to find b.

$$413.34 = \frac{1}{2}b(24.9)$$

$$\frac{413.34 \times 2}{24.9} = b$$

$$33.2 = b$$

By the Pythagorean Theorem, $a^2 + b^2 = c^2$, we know that $c^2 = (24.9)^2 + (33.2)^2 = 620.01 + 1102.24 = 1722.25$, so $c = 41.5$. Adding the values for a, b, and c, we find the perimeter of the triangle is P = 24.9 + 33.2 + 41.5 = 99.6.

(b) From the Pythagorean Theorem, $a^2 + b^2 = c^2$. Thus, we know that $(125.5)^2 + b^2 = (326.3)^2$. Hence, we have $b^2 = 106,471.69 - 15,750.25 = 90721.44$ and $b = 301.2$. The perimeter is P = 125.5 + 301.2 + 326.3 = 753. The area is $A = \frac{1}{2}(301.2)(125.5) = 18,900.3$.

(c) Since A = $\frac{1}{2}ba$, 7518.96 = $\frac{1}{2}$(141.6)a, so

$$a = \frac{7518.96 \times 2}{141.6} = 106.2.$$

By the Pythagorean Theorem, $a^2 + b^2 = c^2$, we know that $(106.2)^2 + (141.6)^2 = c^2$, so $c = 177$. The perimeter is P = 106.2 + 141.6 + 177 = 424.8.

Section 13.2

32. (a) The formula for the area of a trapezoid is

$$A = \frac{1}{2}(a + b) \times h,$$

where a and b are parallel sides, and h is the perpendicular height. We need to find the height of the trapezoid. Construct height h which intersects a base of the trapezoid and divides it into two lengths of 6 units and 3 units. Since the hypotenuse of the right triangle is 5 units long, it must be a 3-4-5 right triangle. Therefore, h must be 4 units long. The area of the trapezoid is A = $\frac{1}{2}$(6 + 9) × 4 = 2 × 15 = 30 square units. The perimeter is P = 4 + 6 + 5 + 9 = 24 units.

(b) The area of the trapezoid is A = $\frac{1}{2}$(6 + 12) × 3 = 27 square units. To find the perimeter we need to find the length of the slanted side of the trapezoid. Since the height segment forms a right angle with the base, and the base angles are congruent, the triangles formed are also congruent. Therefore, the base of each triangle is the same length. Since the base of the trapezoid is 6 units longer than the top, together the triangle bases must be 6 units. Each triangle base must therefore be 3 units long.

By the Pythagorean Theorem, the length of each of the slanted sides of the trapezoid is $3\sqrt{2}$. ($c^2 = 3^2 + 3^2 = 18$, so c = $3\sqrt{2}$.) The perimeter is P = $3\sqrt{2}$ + 6 + $3\sqrt{2}$ + 12 = 18 + $6\sqrt{2}$.

Section 13.2

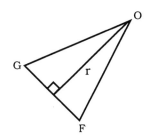

33. (a) Consider ∆OGF. Since r is the height of the triangle, FG is the length of the base, and the area is A = $\frac{1}{2}bh$, we know the area of ∆OGF = $\frac{1}{2}r$ (FG).

(b) All the other triangles are congruent. They differ only in the labeling of the vertices of their bases. The areas of the other triangles are as follows:

$$\Delta OGF, A = \frac{1}{2}r\,(FG) \qquad \Delta OFE, A = \frac{1}{2}r\,(EF)$$

$$\Delta OED, A = \frac{1}{2}r\,(DE) \qquad \Delta ODC, A = \frac{1}{2}r\,(CD)$$

$$\Delta OCB, A = \frac{1}{2}r\,(BC) \qquad \Delta OBA, A = \frac{1}{2}r\,(AB)$$

$$\Delta OAH, A = \frac{1}{2}r\,(HA) \qquad \Delta OHG, A = \frac{1}{2}r\,(GH)$$

Thus, the total area of the regular octagon is given by

$$A = \frac{1}{2}r\,(AB + BC + CD + DE + EF + FG + GH + HA).$$

(c) Since the sum in parentheses is the perimeter of the regular octagon, the formula for the area of a regular polygon is $A = \frac{1}{2}(\text{apothem})(\text{perimeter})$.

Section 13.2

34. The perimeter of any regular n-gon is $n \times s$, where s is the length of a side. The formula from problem 33 part (c) could be written as area $= \frac{1}{2}(\text{apothem})(n \times s) = \frac{1}{2}rns$.

(a) Equilateral triangle: $s = 6$, $r = \sqrt{3}$, $n = 3$.

$$\text{Area} = \frac{1}{2}(\sqrt{3})(6 + 6 + 6) = \frac{1}{2}(\sqrt{3})(6 \times 3) = 9\sqrt{3}.$$

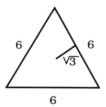

(b) Square: $s = 2\sqrt{2}$, $r = \sqrt{2}$, $n = 4$.

$$\text{Area} = \frac{1}{2}(\sqrt{2})(4)(2\sqrt{2}) = 8.$$

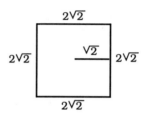

(c) Hexagon: $s = 10$, $r = 5\sqrt{3}$, $n = 6$.

$$\text{Area} = \frac{1}{2}(5\sqrt{3})(6)(10) = 150\sqrt{3}.$$

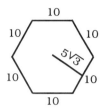

(d) Octagon: $s = 8$, $r = 9.66$, $n = 8$.
    Area $= \frac{1}{2}(9.66)(8)(8) = 309.12$.

(e) 20-gon: $s = 50$, $r = 157.8$, $n = 20$.
    Area $= \frac{1}{2}(157.8)(20)(50) = 78,900$.

(f) 100-gon: $s = 1$, $r = 15.9$, $n = 100$.
    Area $= \frac{1}{2}(15.9)(100)(1) = 795$.

Section 13.2

35. Use the Pythagorean Theorem to check each case. If $a^2 + b^2 = c^2$, then they are the lengths of the sides of a right triangle. Notice that c must always be the largest number.

    (a) $7^2 + 24^2 = 49 + 576 = 625 = 25^2$.
        Yes, forms a right triangle.

    (b) $12^2 + 24^2 = 144 + 576 = 720 \neq 26^2$.
        Not a right triangle.

    (c) $21^2 + 28^2 = 441 + 784 = 1225 = 35^2$.
        Yes, forms a right triangle.

    (d) $11^2 + 60^2 = 121 + 3600 = 3721 = 61^2$.
        Yes, forms a right triangle.

    (e) $8^2 + 9^2 = 64 + 81 = 145 \neq 15^2$.
        Not a right triangle.

    (f) $10^2 + 22^2 = 100 + 484 = 584 \neq 26^2$.
        Not a right triangle.

    Therefore, the numbers given in parts (a), (c), and (d) are lengths of the sides of right triangles.

Section 13.2

36. (a) Without using Pick's Theorem, we could try to divide the polygonal region into polygons whose areas we can calculate. Notice, however, that the region outside the polygon can be divided into triangles and a rectangle. The area of the polygonal region equals the area of square lattice – area outside the polygonal region.

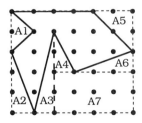

Areas:

Square lattice: A = 5 × 6 = 30.

Triangle 1: A1 = $\frac{1}{2}$(2)(1) = 1.

Triangle 2: A2 = $\frac{1}{2}$(1)(3) = $\frac{3}{2}$.

Triangle 3: A3 = $\frac{1}{2}$(1)(4) = 2.

Triangle 4: A4 = $\frac{1}{2}$(1)(2) = 1.

Triangle 5: A5 = $\frac{1}{2}$(2)(2) = 2.

Triangle 6: A6 = $\frac{1}{2}$(3)(1) = $\frac{3}{2}$.

Rectangle: A7 = 4 × 2 = 8.

The area of the region outside the polygon = 1 + $\frac{3}{2}$ + 2 + 1 + 2 + $\frac{3}{2}$ + 8 = 17. Therefore, the area of the polygonal region = 30 − 17 = 13 square units.

(b) Using Pick's Theorem, $A = \left(\frac{b}{2} + i - 1\right)$, find b and i in each case.

(i) b = number of dots on the border = 14. i = number of dots on the inside = 0. $A = \left(\frac{b}{2} + i - 1\right) = \frac{14}{2} + 0 - 1 = 7 - 1 = 6$ square units.

(ii) b = number of dots on the border = 35. i = number of dots on the inside = 68. $A = \left(\frac{b}{2} + i - 1\right) = \frac{35}{2} + 68 - 1 = 84.5$ square units.

Section 13.2

37. Calculate each of the lengths.

(1) The area of a square is A = s^2, where s = side length, A = 100 = s^2, so s = 10. The perimeter is 4(10) = 40 units.

(2) The formula for the circumference of a circle is C = $2\pi r$. Therefore, the circumference is C = $2\pi(5)$ = $10\pi \approx 31.4$ units.

(3) The perimeter of a triangle is the sum of the lengths of the three sides. P = 10 + 10 + 9 = 29 units.

In order from smallest to largest, the lengths are the perimeter of the triangle, the circumference of the circle, and the perimeter of the square.

Section 13.2

38. Let w = the width of the rectangle. The length is $w + 3$ since the length is 3 cm more than its width. Since the area of a rectangle is length $\times$ width, we know

$$40 = (w + 3)w$$
$$0 = w^2 + 3w - 40$$
$$0 = (w - 5)(w + 8)$$
$$w = 5 \text{ or } w = -8$$

Since width cannot be negative we eliminate $w = -8$. Therefore, the width is 5 cm and the length is $5 + 3 = 8$ cm.

Section 13.2

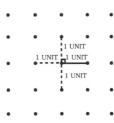

39. Consider any triangle with one side horizontal. Let the horizontal distance between two adjacent dots be 1 unit. In an equilateral triangle, all three sides must be the same length. If the horizontal side has length 1, then the other two sides must each have length 1 also. Notice, however, that the only other dots that are 1 unit away lie directly above, directly below, or along the same line as the endpoints of the original side. Since the dots directly above or directly below form right angles when connected to the horizontal base, and since a triangle cannot be constructed from three collinear points, no equilateral triangle can be constructed having a horizontal side. (A similar argument holds for a triangle having a vertical side.) If we are to construct an equilateral triangle on the square lattice, then it must have no sides horizontal or vertical. Through trial and error, we can see that we can almost always find two sides of equal length, but the third side will never be the same length. Therefore, no equilateral triangle can be constructed on the square lattice.

Section 13.2

40. Building square corners requires a method for constructing right angles. Every right triangle has one right angle, and we know there are right triangles with predictable leg lengths as in a 3-4-5 right triangle. The rope is 12 units long. Since $3 + 4 + 5 = 12$, the rope can be used to build a right triangle with the right angle at the square corner.

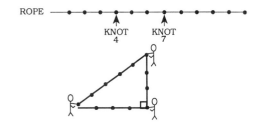

Have 3 Egyptians hold the rope. One holds the two loose ends of the rope. One holds the rope at the fourth knot, and the other holds the rope at the seventh knot. Have the Egyptians pull the rope taut. When the ends of the rope just touch, the right triangle has been formed.

Section 13.2

41. Check to see if the Pythagorean Theorem holds for each set of lengths. If it does not hold, then the triangle is either acute or obtuse. Consider an equilateral triangle which has three 60° angles. It is also called an acute triangle. Suppose its sides were each of length 4; that is, let $a = 4$, $b = 4$, and $c = 4$. Then $a^2 + b^2 = 4^2 + 4^2 = 16 + 16 = 32$, and $c^2 = 4^2 = 16$. Therefore, for an acute triangle, $a^2 + b^2 > c^2$. For an obtuse triangle, then, $a^2 + b^2 < c^2$, where c is the length of the longest side.

 (a) $a^2 + b^2 = 54^2 + 70^2 = 2916 + 4900 = 7816 < 8100 = 90^2 = c^2$. The triangle is obtuse.

 (b) $a^2 + b^2 = 16^2 + 63^2 = 256 + 3969 = 4225 = 65^2 = c^2$. The triangle is a right triangle.

 (c) $a^2 + b^2 = 24^2 + 48^2 = 576 + 2304 = 2880 > 2704 = 52^2 = c^2$. The triangle is acute.

 (d) $a^2 + b^2 = 27^2 + 36^2 = 729 + 1296 = 2025 = 45^2 = c^2$. The triangle is a right triangle.

 (e) $a^2 + b^2 = 46^2 + 48^2 = 2116 + 2304 = 4420 > 2500 = 50^2 = c^2$. The triangle is acute.

 (f) $a^2 + b^2 = 9^2 + 40^2 = 81 + 1600 = 1681 < 2116 = 46^2 = c^2$. The triangle is obtuse.

Section 13.2

42. (a) To form a triangle, any two sides combined must be longer than the third side. Since c is the longest side, if $a + b > c$, then a triangle can be formed.

 (b) From the Pythagorean Theorem, if $a^2 + b^2 = c^2$ or $c = \sqrt{a^2 + b^2}$, then the triangle will be a right triangle.

 (c) From problem 41, a triangle is an acute triangle if $a^2 + b^2 > c^2$, or $c < \sqrt{a^2 + b^2}$.

 (d) From problem 41, a triangle is an obtuse triangle if $a^2 + b^2 < c^2$, or $c > \sqrt{a^2 + b^2}$.

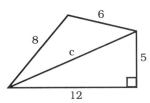

Section 13.2

43.   Recall that Hero's formula tells us how to find the area of a triangle if we know the lengths of three sides. We want to build a right triangle in the given quadrilateral so we can apply Hero's formula. Construct the diagonal from the lower left vertex to the upper right vertex. Since a right triangle is formed, by the Pythagorean Theorem, the diagonal has length $c = \sqrt{a^2 + b^2} = \sqrt{5^2 + 12^2} = 13$. Now we have two triangles, and we know all the side lengths. Find the area of each triangle and add to find the area of the quadrilateral.

Hero's formula: Area = $\sqrt{s(s - a)(s - b)(s - c)}$,

where a, b, and c are lengths of the sides and $s = \dfrac{a + b + c}{2}$.

$$
\begin{aligned}
\text{Area of right triangle} &= \sqrt{15(15 - 5)(15 - 12)(15 - 13)} \\
&= \sqrt{15(10)(3)(2)} \\
&= 30 \text{ square units.}
\end{aligned}
$$

$$
\begin{aligned}
\text{Area of other triangle} &= \sqrt{13.5(13.5 - 8)(13.5 - 6)(13.5 - 13)} \\
&= \sqrt{13.5(5.5)(7.5)(0.5)} \\
&\approx 16.69 \text{ square units.}
\end{aligned}
$$

Area of quadrilateral ≈ 30 + 16.69 ≈ 46.69 square units.

Section 13.2

44.   (a) In the diagram, since there are parallel lines cut by a transversal, the measure of the indicated angle is $7\dfrac{1}{2}$ degrees by the Alternate Interior Angles Congruence Property.

(b) Set up a proportion using the ratios of arc length to central angle:

$$
\frac{500 \text{ miles}}{7.5 \text{ degrees}} = \frac{\text{circumference (in miles)}}{360 \text{ degrees}}
$$
$$
24{,}000 \text{ miles} = c
$$

(c) Eratosthenes was off by only 24,901.55 − 24,000 = 901.55 miles.

Section 13.2

45. The diameter of the hole is also the diagonal of the square plug. Recall that a square has equal side lengths. Let $x =$ the length of the side of the square. Since the square has all right angles, we can find x using the Pythagorean Theorem.

$$x^2 + x^2 = (3.16)^2$$
$$2x^2 = 9.9856$$
$$x^2 = 4.9928$$
$$x \approx 2.23 \text{ cm.}$$

Therefore, the square must have a side length of about 2.23 cm.

Section 13.2

46. Since the segments $\overline{OA}$ and $\overline{AB}$ are perpendicular, right angles are formed. Length $\overline{AB}$ is 1 unit. Let $x =$ the length of $\overline{OA}$. By the Pythagorean Theorem, the length of $\overline{OB}$ can be calculated as follows: length of $\overline{OB} = \sqrt{x^2 + 1^2} = \sqrt{x^2 + 1}$.

Therefore, the radius of the outer circle is $\sqrt{x^2 + 1}$, and the radius of the inner circle is x. If we calculate the area of the large circle and subtract the area of the small circle, we will be left with the area in between. Recall that the area of a circle $= \pi r^2$.

Area of large circle $= \pi(\sqrt{x^2 + 1})^2 = \pi(x^2 + 1) = \pi x^2 + \pi$.
Area of small circle $= \pi(x^2) = \pi x^2$.
Area in between $= \pi x^2 + \pi - \pi x^2 = \pi$ square units.

Section 13.2

47. Recall that the area of a circle is πr^2. It follows that the area of a semicircle is $\dfrac{\pi r^2}{2}$. Calculate the area of each semicircle. Notice that the radius of each semicircle is 4. The area of the upper semicircle is $\dfrac{\pi(4)^2}{2} = \dfrac{16\pi}{2} = 8\pi$. The area of the left hand semicircle is $\dfrac{\pi(4)^2}{2} = \dfrac{16\pi}{2} = 8\pi$. Notice that the "petal" is included in both calculations and has been counted twice. The other two semicircles will each be 8π square units in area. If we shade the appropriate regions, we will see that the area of each of the petals will be counted twice. Therefore, the area of the petal can be found by finding the area of the four semicircles and subtracting the area of the square.

Area of petals = areas of four semicircles – area of square
$$= 8\pi + 8\pi + 8\pi + 8\pi - 64$$
$$= 32\pi - 64 \text{ square units.}$$

Section 13.2

48. (a) The field measures (100)(100) = 10,000 m². The diameter of the circle is 100 m so the radius is 50 m. The area of irrigation is $\pi(50)^2$ = 2500π m². The percent irrigated is $\frac{2500\pi}{10000} \approx 0.785 = 78.5\%$.

    (b) Since the field measures 100 m on a side and two circles span a side, the diameter of each circle is 50 m. The radius of each circle is 25 m, and the area of each is $\pi(25)^2$ = 625π m². The total area of irrigation is 4(625π m²) = 2500π m². The percent irrigated is $\frac{2500\pi}{10000} \approx 0.785 = 78.5\%$.

    (c) Both systems will irrigate the same amount of land.

    (d) Since they both irrigate the same amount, it is unnecessary to use more sprinklers with a smaller radius.

Section 13.2

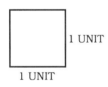

1 UNIT

1 UNIT

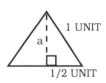

1 UNIT

a

1/2 UNIT

1 UNIT

49. The figure is composed of 1 regular hexagon, 6 squares, and 6 equilateral triangles. Find the areas of each and add to find the total area of the figure. Since the length of the side of a triangle is 1 unit, the length of the sides of the squares and hexagon are 1 unit.

Area of 6 squares = 6 × (area of 1 square) = 6(1)² = 6 square units. Area of 6 triangles = 6 × (area of 1 triangle) =

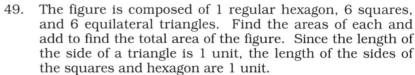

$6 \times \frac{1}{2} \times 1 \times \frac{\sqrt{3}}{2} = \frac{3\sqrt{3}}{2}$ square units. Recall that a hexagon can be divided into 6 equilateral triangles, each with a side length of 1 unit.

Area of hexagon = 6 × (area of 1 triangle)

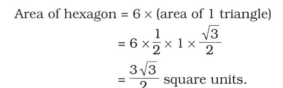

$$= 6 \times \frac{1}{2} \times 1 \times \frac{\sqrt{3}}{2}$$

$$= \frac{3\sqrt{3}}{2} \text{ square units.}$$

Therefore, the area of the figure = $6 + \frac{\sqrt{3}}{2} + \frac{\sqrt{3}}{2} = 6 + 3\sqrt{3}$ square units.

Section 13.2
50. To travel from Portland to the Aral Sea over the North Pole, we would travel over an arc with central angle 90° or one-fourth of the circumference of the earth. Since the circumference is $2\pi r$, and $r = 6380$ km, the circumference is $2\pi(6380)$ km. We would travel $\frac{1}{4}(2\pi)(6380)$ or $3190\pi \approx$ 10,021.7 km. The triangle formed from the vertices at the center of the earth, Portland, and the point where the horizontal line through Portland intersects the North-South line is a 45-45-90 degree triangle. Since we know the hypotenuse, the side lengths are $3190\sqrt{2}$, by the Pythagorean Theorem. We want half of the circumference of the circle through the 45th parallel.

The circumference is $2\pi(3190\sqrt{2})$ km. Half of the circumference is $\pi(3190\sqrt{2}) \approx 14{,}172.8$ km. It would be shorter to travel over the North Pole.

Section 13.2

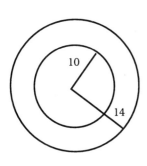

51. (a) Consider the two concentric circles. Area of the small circle = $\pi(10)^2 = 100\pi$. Area of large circle = $\pi(14)^2 = 196\pi$. Area between the circles = $196\pi - 100\pi = 96\pi$ square units.
   (b) Let x = the radius of the large circle. If the area of inner circle = area between circles, then
$$100\pi = \pi x^2 - 100\pi$$
$$200\pi = \pi x^2$$
$$200 = x^2$$
$$10\sqrt{2} = x.$$

Therefore, the radius of the larger circle must be $10\sqrt{2}$ square units.

Section 13.2
52. Let x = the width (radius) of the innermost ring. The radius of the inner shaded ring is $3x$. The outer shaded ring is formed by taking the whole region, which has a radius of $5x$, and removing the inner 4 rings, which have a combined radius of $4x$. Recall that the area of a circle is πr^2.

Area of inner shaded ring: $\pi(3x)^2 = 9x^2\pi$.
Area of outer shaded ring:
$$\pi(5x)^2 - \pi(4x)^2 = 25x^2\pi - 16x^2\pi = 9x^2\pi.$$

The areas are the same.

Section 13.3

8. (a) Divide each hexagonal base into six equilateral triangles by connecting opposite vertices.

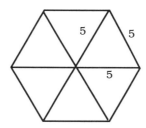 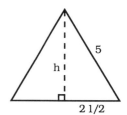

Since the side length of the hexagon is 5 units, the leg lengths for the equilateral triangles will all be 5 units. Area of the hexagon = 6(area of an equilateral triangle). Consider one equilateral triangle. Construct the altitude, or height. By the Pythagorean Theorem, we know that the altitude can be found by solving the following equation.

$$h^2 + \left(2\frac{1}{2}\right)^2 = 5^2$$

$$h^2 = 25 - 6\frac{1}{4} = 18\frac{3}{4}$$

$$h = \frac{5\sqrt{3}}{2}.$$

The area of one of the equilateral triangles is equal to $\frac{1}{2}$ base × height = $\frac{1}{2}(5)\frac{5\sqrt{3}}{2}$ = $\frac{25\sqrt{3}}{4}$ square units.

Therefore, the area of the hexagon is $6 \times \frac{25\sqrt{3}}{4}$ = $\frac{75\sqrt{3}}{2}$ square units.

(b) The total surface area = area of two hexagonal bases + area of six rectangular sides. Since the area of a rectangle is length × width, the area of each rectangle is 10 × 5 = 50 square units. Surface area = $2\left(\frac{75\sqrt{3}}{2}\right) + 6(50) = 75\sqrt{3} + 300$ square units.

Section 13.3

9. Consider a box. Let l = length, w = width, and h = height. The new box will have dimensions $2l$, $2w$, and $2h$. Calculate the surface area for each box and compare.

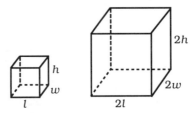

Each box has six sides. The surface area is the sum of the areas of the six sides.

Surface area for box 1 = $2l \times w + 2w \times h + 2l \times h$.

Surface area for box 2 = $2(2l \times 2w) + 2(2w \times 2h) + 2(2l \times 2h)$.
$$= 8l \times w + 8w \times h + 8l \times h$$
$$= 4(2l \times w + 2w \times h + 2l \times h)$$
$$= 4(\text{area for box 1}).$$

Notice that box 2 requires 4 times as much cardboard as does box 1.

Section 13.3

10. (a) Any right rectangular prism has a surface area found by calculating $2lw + 2wh + 2lh$. Use systematic guess and test to find arrangements of 36 cubes that yield the required surface area, 96.

L	W	H	SA
36	1	1	72 + 2 + 72 = 146
18	2	1	72 + 4 + 36 = 112
12	3	1	72 + 6 + 24 = 102
9	4	1	72 + 8 + 18 = 98
6	6	1	72 + 12 + 12 = 96

The arrangement is a right rectangular prism with dimensions 6 by 6 by 1 unit.

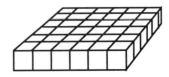

(b) Continue adjusting the dimensions from part (a). Notice that we cannot, however, adjust only length and width. Consider a height of 2 cubes.

L	W	H	SA
9	2	2	36 + 8 + 36 = 80

The required arrangement has dimensions 9 by 2 by 2.

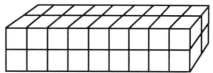

(c) Notice that as we increase the height and bring the length and width closer in dimension, we decrease the surface area.

L	W	H	SA
6	3	2	36 + 12 + 24 = 72
4	3	3	24 + 18 + 24 = 66

Since there are no other unique dimension combinations, the dimensions 4 × 3 × 3 give the smallest surface area of 66 square units.

(d) Notice that when the dimensions were not similar, the surface area was larger. From part (a), we see that when the difference in dimensions is the most extreme, (36 × 1 × 1), the largest surface area of 146 square units was obtained.

Section 13.3
11. Since the scale is 5 cm = 3 m, square both sides to find the scale for area: 25 cm^2 = 9 m^2. Use dimensional analysis to find the area of the exposed surfaces of the building in square meters. We can use the surface area of the model to determine the surface area of the building.

$$27{,}900 \text{ cm}^2 \times \frac{9\text{m}^2}{25\text{cm}^2} = 10{,}044 \text{ m}^2$$

The area of the exposed surface of the building is 10,044 m^2.

Section 13.3
12. Draw a picture. Let w = width, h = height, and l = length. Set up three area equations:

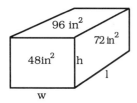

Area of end: 48 = wh.
Area of side: 72 = lh.
Area of top: 96 = lw.

Substitute and solve:

$$w = \frac{48}{h}, \quad \text{so} \quad 96 = l\left(\frac{48}{h}\right)$$

$$l = \frac{72}{h}, \quad \text{so} \quad 96 = \frac{72}{h} \times \frac{48}{h}$$

$$96 = \frac{3456}{h^2}$$

$$h^2 = \frac{3456}{96}.$$

$$h^2 = 36, \quad \text{so} \quad h = 6.$$

Since $h = 6$ in., $l = \frac{72}{6} = 12$ in. and $w = \frac{48}{6} = 8$ in., the dimensions of the box are 12 inches by 8 inches by 6 inches.

Section 13.3

13. (a) Since the earth has a diameter of 12,760 kilometers, it has a radius of $\frac{12,760}{2} = 6380$ kilometers.

   (b) Recall the formula for the surface area of a sphere: $S = 4\pi r^2$. Therefore, the surface area of the earth is $4\pi(6380)^2 = 4\pi(40,704,400) \approx 5.115 \times 10^8$ km^2.

   (c) Since land area is 135,781,867 km^2, the percent of the Earth's surface that is land can be found by taking $\frac{\text{land area}}{\text{total surface area}} \times 100\% = \frac{135781867}{5.12 \times 10^8} \times 100\% \approx 26.5\%$.
   Therefore, about 26.5% of the earth's surface is land.

Section 13.3

14. Recall that the surface area of a sphere is $4\pi r^2$ where r is the radius. Consider sphere 1 with radius r. If the radius is cut in half, the radius for sphere 2 will be $\frac{r}{2}$.

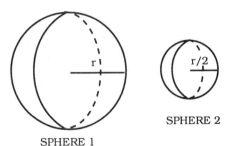

SPHERE 1

SPHERE 2

Surface area (sphere 1) = $4\pi r^2$.

Surface area (sphere 2) = $4\pi \left(\dfrac{r}{2}\right)^2 = 4\pi\,\dfrac{r^2}{4} = \pi r^2$.

Therefore, when the radius of a sphere is reduced by half, the surface area is one-fourth as large.

Section 13.3

15. A square is rolled up to form a cylinder. The surface area can be found by adding the lateral surface (the original square) and the circular bases.

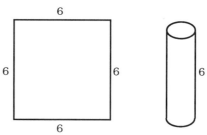

Since the circumference of a circle is $2\pi r$, and the circle has a circumference of length 6 cm, $6 = 2\pi r$, so $r = \dfrac{6}{2\pi} = \dfrac{3}{\pi}$ cm. The area of each of the circular bases is $\pi r^2 = \pi\left(\dfrac{3}{\pi}\right)^2 = \dfrac{9}{\pi}$ cm². The total area of the cylinder is $\dfrac{9}{\pi} + \dfrac{9}{\pi} + 36 = \dfrac{18}{\pi} + 36$ cm².

Section 13.3

16. Notice that the sphere has a diameter of 10 units, so its radius is 5 units. The smallest cylinder that would contain the sphere would be such that the top, bottom, and sides touch the sphere. The height of the cylinder would be the diameter of the sphere. The radius of the base of the cylinder would be the radius of the sphere.

Therefore, the surface area of the smallest cylinder that contains the sphere is $2\pi r^2 + 2\pi rh = 2\pi(5)^2 + 2\pi(5)(10) = 50\pi + 100\pi = 150\pi$ square units.

Section 13.3

17. Construct a model. Notice that the unfolded cylinder is a rectangle with height = 1 meter and width = circumference of a circle of radius 10 cm. (Recall that 1 meter = 100 cm.)

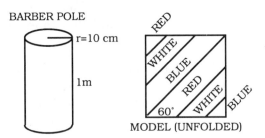

Since there are only 3 stripes on the pole, each stripe must cover $\frac{1}{3}$ of the circumference at the base of the pole.

Unfolded, each stripe would cover $\frac{1}{3}$ of the width of the rectangle. Also, each stripe makes a constant angle of 60° with the vertical. If we cut out the pieces and put each color together, the areas covered by each color are equal.

Each stripe covers $\frac{1}{3}$ of the total surface area.

Surface area is $2\pi rh = 2\pi(10)(100) = 2000\pi$ cm^2.

Red covers $\frac{1}{3}$ surface area = $\frac{1}{3}(2000\pi) \approx 2094$ cm^2.

Section 13.4

5. Since they are stacks of unit cubes, each cube has a volume of 1 cubic unit. Therefore the volume of each stack is the total number of cubes in the stack. When calculating the surface area for each stack, consider each of the six sides carefully. It may be helpful to sketch each of the side views. We will denote the sides as "Top", "Bottom", "Front", "Back", Right" , and "Left". (Note: If you have a difficult time visualizing each side, try using dice to build each stack.)

   (a) The volume is the number of cubes: 3 + 4 + 2 + 1 + 1 + 3 = 14 cubic units. Consider the following sketches when finding the surface area.

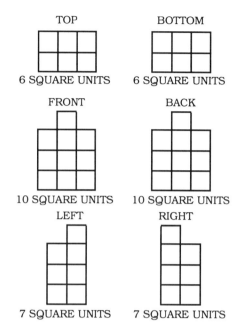

The surface area is the sum of the areas of the sides. Surface area = 6 + 6 + 10 + 10 + 7 + 7 = 46 square units. Notice that the top and bottom views are the same. The front and back views are the same, and the right and left views are the same. When calculating surface area, we need only find the areas of the top, front, and left side and multiply by two.

(b) The volume is 1 + 4 + 2 = 7 cubic units. For surface area, consider the top, front, and left side views.

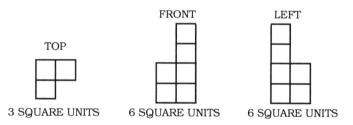

Surface area = 2(3 + 6 + 6) = 2(15) = 30 square units.

(c) The volume is 3 + 3 + 3 + 1 + 2 + 3 + 1 + 2 + 3 = 21 cubic units. For surface area, consider the top, front, and left side views.

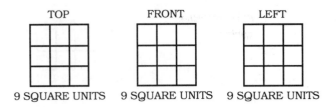

TOP   FRONT   LEFT

9 SQUARE UNITS   9 SQUARE UNITS   9 SQUARE UNITS

Surface area = 2(9 + 9 + 9) = 54 square units.

Section 13.4

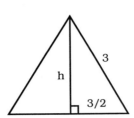

6. Each shape has a "cut" made in it. We will find the volume of the shape without the "cut", then subtract the volume of the "cut" to find the volume of the shape.

(a) Without the "cut", the figure is a rectangular prism with length = 9 inches, width = 8 inches, and height = 8 inches. Since the volume of a rectangular prism is length × width × height, the volume without the "cut" is 9 × 8 × 8 = 576 cubic inches. The "cut" is a triangular prism with an equilateral triangular base which has side lengths of 3 inches each. Since the volume of a prism is the product of the area of the bases and the height, we need to find the area of the base. The area of a triangle is $\frac{1}{2} bh$. By the Pythagorean Theorem,

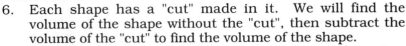

$$h = \sqrt{3^2 - \left(\frac{3}{2}\right)^2} = \frac{3\sqrt{3}}{2}$$

$$\text{Area} = \frac{1}{2}(3)\left(\frac{3\sqrt{3}}{2}\right) = 9\frac{\sqrt{3}}{4} \text{ square inches.}$$

Volume of the triangular prism $= \left(\frac{9\sqrt{3}}{4}\right)(8) = 18\sqrt{3} \text{ in}^3$.

Therefore, the volume of the figure = volume without "cut" – volume of "cut".

Volume $= 576 - 18\sqrt{3} \approx 545$ cubic inches.

(b) Without the "cut", the figure is a hemisphere with radius 4.5 inches (diameter 9 inches). The "cut" is a hemisphere with radius 3.5 inches. Find the volume of each hemisphere, and then subtract to find the volume of the figure. (Note: The volume of a sphere is $\frac{4}{3}\pi r^3$, so the volume of a hemisphere is $\frac{2}{3}\pi r^3$.) The volume of the larger hemisphere is $\frac{2}{3}\pi r^3 =$ $\frac{2}{3}\pi(4.5)^3 \approx 60.75\pi \text{ in}^3$. The volume of the smaller hemisphere is $\frac{2}{3}\pi r^3 = \frac{2}{3}\pi(3.5)^3 = 28.58\pi$ in^3. Thus, the volume of the figure is $60.75\pi - 28.58\pi \approx 101.05$ cubic inches.

Section 13.4

7. (a) Calculate the area of each side.

SHALLOW END WALL: This wall is a 1 meter by 20 meter rectangle, so the area is $1 \times 20 = 20$ square meters.
SLANTED FLOOR: This is a 13 meter by 20 meter rectangle, so the area is $13 \times 20 = 260$ square meters.
FLAT BOTTOM: This is a 13 meter by 20 meter rectangle, so the area is $13 \times 20 = 260$ square meters.
DEEP END WALL: This wall is a 6 meter by 20 meter rectangle, so the area is $6 \times 20 = 120$ square meters.
TWO SIDE WALLS: If we construct a vertical line at the point where the flat bottom and slanted floor meet, we will divide each side into two pieces, a rectangle and a trapezoid.

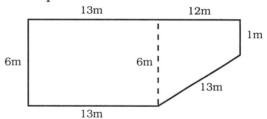

The area of each side = area of a rectangle + area of a trapezoid = $(6 \times 13) + \frac{1}{2}(12)(1 + 6) = 78 + 42 = 120$.

Therefore, the area of the two sides combined is $2(120) = 240$ square meters. We will need $20 + 260 + 260 + 120 + 240 = 900$ square meters of tile.

(b) Notice that the pool is a prism with an irregularly shaped base. The volume of a prism is the product of the area of the base and the height. We calculated the area of the base (area of the side of the pool) in part (a), and found it to be 120 square meters. The height is 20 meters. Therefore, the volume of the pool is $120(20) = 2400$ cubic meters.

Section 13.4

8. Suppose that the tennis balls touch the sides, the top, and the bottom of the can. Let r = the radius of a tennis ball.
    (a) The circumference of the can is $2\pi r$. The height is $6r$. Since $2\pi > 6$, circumference > height.
    (b) The percent of the can occupied by air =

$$\frac{\text{volume of the can} - \text{volume of 3 balls}}{\text{volume of can}} \times 100\%.$$

Volume of the can = $\pi r^2 h = \pi(3.5)^2(21) = 257.25\pi$ cm^3.

Volume of 3 balls = 3(volume of 1 ball) = $3\left(\dfrac{4}{3}\pi r^3\right) =$

$4\pi \times (3.5)^3 = 171.5\pi$ cm^3. The percent occupied by air =
$\dfrac{257.25\pi - 171.5\pi}{257.25\pi} \times 100\% = \dfrac{85.75\pi}{257.25\pi} \times 100\% = 33\dfrac{1}{3}\%.$

Section 13.4

9. Draw a picture for each problem.

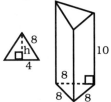

(a) The bases of the prism are equilateral triangles. Volume = area of triangular base × height of prism.

The altitude of triangle is given by $h = \sqrt{8^2 - 4^2} =$

$4\sqrt{3}$. The area of each triangular base is $\dfrac{1}{2}bh =$

$\dfrac{1}{2}(8)(4\sqrt{3}) = 16\sqrt{3}$. The height of the prism is 10.

Therefore, the volume is $16\sqrt{3} \times 10 = 160\sqrt{3}$ cubic units.

(b) The bases of the prism are trapezoids. The volume of the prism = area of trapezoidal base × height of prism.

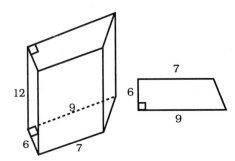

The area of the trapezoid is $\frac{1}{2}(b_1 + b_2)h = \frac{1}{2}(7 + 9)6 = 48$. The height of the prism is 12. Therefore, the volume of the prism is $(48)(12) = 576$ cubic units.

(c) The bases of the prism are right triangles. The volume of the prism = area of triangular base × height of prism.

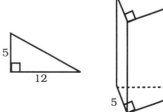

The area of the right triangle $= \frac{1}{2}bh = \frac{1}{2}(12)(5) = 30$.

The height of the prism is 20. Therefore, the volume of the prism is $(30)(20) = 600$ cubic units.

Section 13.4

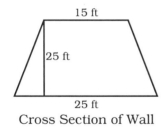

15 ft

25 ft

25 ft

Cross Section of Wall

10. Think of the Great Wall as a prism. The volume of a prism is the area of the base × height. The base is a trapezoid. The "height" is the length of the wall, which is 1500 miles.

The area of the trapezoidal base $= \frac{1}{2}(25)(15 + 25) = 500$ square feet. The height of the prism = 1500 miles × $\frac{5280 \text{ feet}}{1 \text{ mile}} = 7{,}920{,}000$ feet. Therefore, the volume of the wall is $(500)(7{,}920{,}000) = 3{,}860{,}000{,}000$ cubic feet. Recall that 3 feet = 1 yard, so 27 cubic feet = 1 cubic yard.

$3{,}960{,}000{,}000 \text{ ft}^3 \times \frac{1 \text{ yd}^3}{27 \text{ ft}^3} \approx 1.47 \times 10^8 \text{ yd}^3$. Therefore, about 1.47×10^8 cubic yards of material make up the Great Wall of China.

Section 13.4

11. We need to find the volume of the cylinder whose diameter is 2 feet (radius 1 foot) and whose height is 3 feet.
Volume $= \pi r^2 h = \pi(1)^2(3) = 3\pi$ square feet.

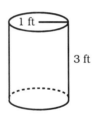

1 ft

3 ft

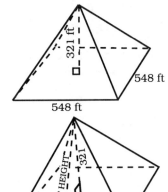

548 ft

Section 13.4

12. (a) The volume of a pyramid is $\frac{1}{3}$ (area of base) (height). The base is a square with side lengths 548 feet. The area of the square = 548^2 = 300,304 square feet. Therefore, the volume = $\frac{1}{3}$(300304) (321) = 32,132,528 ft^3, or about 32,100,000 ft^3 .

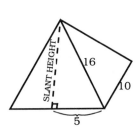

(b) The lateral surface area excludes the base and is calculated by $\frac{1}{2}$ × (perimeter) × (slant height). The perimeter is (4)(548) = 2192 feet. By the Pythagorean Theorem, we calculate the slant height to be $\sqrt{(321)^2 + (274)^2}$ ≈ 422.04 feet. The lateral surface area is $\frac{1}{2}$ (2192)(422.04) ≈ 462,555 or about 463,000 square feet.

Section 13.4

13. (a) When folded, the net becomes a square pyramid. The volume is $\frac{1}{3}$ (area of the base) × (height). The area of the square base is 10 × 10 = 100 square centimeters. To calculate the height, we will need the slant height. By the Pythagorean theorem, the slant height is $\sqrt{16^2 - 5^2} = \sqrt{231}$. The height of the pyramid is given by $\sqrt{16^2 - 5^2} = \sqrt{231 - 25} = \sqrt{206}$. Therefore, the volume is $\frac{1}{3}$(100) $\sqrt{206}$ ≈ 478 cubic centimeters. The surface area if the sum of the four triangular sides and the square base. Surface area = $4\left[\frac{1}{2}(10)\sqrt{231}\right] + (10)(10) = 20\sqrt{231} + 100 ≈ 404$ cm^2.

(b) When folded, the net becomes a hexagonal prism. The volume of any prism is (area of the base) × (height). The area of the hexagonal base can be found by dividing the region into six equilateral triangles. The area of one triangle is $\frac{1}{2}bh$.

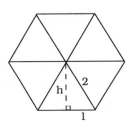

By the Pythagorean Theorem, $h = \sqrt{2^2 - 1^2} = \sqrt{3}$, so the area of one triangle is $\frac{1}{2}(2)\sqrt{3} = \sqrt{3}$. Therefore, the area of the hexagonal base is $6\sqrt{3}$. The volume is $(6\sqrt{3})(12) = 72\sqrt{3} \approx 125$ cubic centimeters. The total surface area is the sum of the areas of the two hexagonal bases and the areas of the six rectangular sides. Surface area $= 2(6\sqrt{3}) = 6(2)(12) = 12\sqrt{3} + 144 \approx 165$ square centimeters.

Section 13.4

14. Since the scale is 5 cm = 3 m, cube both sides to find the scale for volume: 125 cm^3 = 27 m^3. Use dimensional analysis to find the volume of the finished structure in cubic meters. 396,000 cm$^3 \times \dfrac{27 \text{ m}^3}{125 \text{ cm}^3} = 85{,}536$ m^3. The volume of the finished structure is 85,536 m^3.

Section 13.4

15. (a) Divide the regular hexagon into 6 equilateral triangles each having side length 10 feet, by connecting 3 pairs of opposite vertices. To find the area of the triangle, we need to find the altitude. By the Pythagorean Theorem, $5^2 + h^2 = 10^2$, so $h = \sqrt{100 - 25} = 5\sqrt{3}$. Therefore, the area of each equilateral triangle is $\frac{1}{2}bh = \frac{1}{2}(10)(5\sqrt{3}) = 25\sqrt{3}$. The area of the hexagon is 6 × (area of one triangle) $= 6(25\sqrt{3}) = 150\sqrt{3}$ square feet.

    (b) If the patio is 4 inches thick, then we have a right prism with hexagon bases. Volume of a prism = Ah. A = area of the hexagon, and h = height of the prism. We calculated the area of the hexagon in part (a). The area is $150\sqrt{3}$ square feet. The height is 4 inches. Before we find the volume we need consistent units. Since 4 inches $= \frac{1}{3}$ foot, the volume $= (150\sqrt{3})(\frac{1}{3}) = 50\sqrt{3}$ cubic feet.

    (c) From part (b), we know that the volume of the patio is $50\sqrt{3}$ cubic feet. If the concrete costs $45 per cubic yard, then we need to convert the volume from cubic feet to cubic yards. Use dimensional analysis. (Note: 3 feet = 1 yard, so 27 feet^3 = 1 yard^3.) The cost of the concrete is $50\sqrt{3}$ feet$^3 \times \dfrac{1 \text{ yard}^3}{27 \text{ feet}^3} \times \dfrac{\$45}{1 \text{ yard}^3} \approx \$144.34$. The concrete will cost $144.34.

Section 13.4

16. (a) First find the volume of the spherical tank. If the diameter is 60 feet, then the radius is 30 feet. The volume of a sphere is $\frac{4}{3}\pi r^3$. Volume $= \frac{4}{3}\pi(30)^3 = 36,000\pi$ ft^3. The formula for the volume of a right circular cylinder is $\pi r^2 h$. We want the volume of the cylinder to be the same as the spherical tank, 36000π, and we know the radius of the cylinder is 30 feet.

$$\pi(30)^2 h = 36,000\pi$$
$$900\pi h = 36000\pi$$
$$h = 40 \text{ feet.}$$

(b) Each tank has a volume of $36,000\pi$ cubic feet. Use dimensional analysis to change cubic feet to gallons.

$$36,000\pi \text{ cubic feet} \times \frac{7.5 \text{ gallons}}{1 \text{ cubic foot}} \approx 848,000 \text{ gallons.}$$

(c) The formula for the surface area of a sphere is $4\pi r^2$. Since the radius of the sphere is 30 feet, the surface area is $4\pi(30)^2 = 3600\pi$ square feet. The formula for the surface area of a cylinder is $2\pi r^2 + 2\pi r h$. Since the radius of the cylinder is 30 feet and the height is 40 feet, the surface area is $2\pi(30)^2 + 2\pi(30) \times (40) = 1800\pi + 2400\pi = 4200\pi$ square feet. Therefore, the sphere will require less material in its construction.

Section 13.4

17. Calculate the volume of the sculpture in cubic centimeters, keeping in mind that the sphere and the square prism are both hollow. They each have a thickness of 2 mm, or 0.2 cm. Consider each piece separately.

Sphere: Since the outer radius is 18 cm, and the sphere is 0.2 cm thick, the inner radius is 17.8 cm. The volume of metal can be found by calculating the volume of a sphere with radius 18 cm and subtracting the volume of a sphere with radius 17.8 cm. The volume of the sphere sculpture $= \frac{4}{3}\pi(18)^3 - \frac{4}{3}\pi(17.8)^3 \approx 7776\pi - 7519.7\pi \approx 256.3\pi \approx 805.29$ cm^3.

Prism Since the outside dimensions are 40 cm by 40 cm by 1 m (100 cm) thick, the inside dimensions are 39.6 cm by 39.6 cm by 99.6 cm. The volume of the metal can be found by calculating the volume of the prism with the "outside" dimensions and subtracting the volume of the prism with the "inside" dimensions. The volume of the prism sculpture $= (40)(40)(100) - (39.6)(39.6)(99.6) = 160,000 - 156,188.736 \approx 3811.26$ cm^3. The total volume of the metal in the sculpture is $805.29 + 3811.26 \approx 4616.55$ cm^3. Since the density of iron is 7.87 g/cm^3, we can find the weight by multiplying the volume by the density of iron. The weight of the sculpture is

$$4616.56 \text{ cm}^3 \times 7.87 \frac{\text{g}}{\text{cm}^3} \times \frac{1\text{kg}}{1000\text{g}} \approx 36.3 \text{ kg}.$$

The sculpture weighs approximately 36.3 kilograms.

Section 13.4
18. (a) The volume of the rock is equal to the volume of the water it displaces in the aquarium.

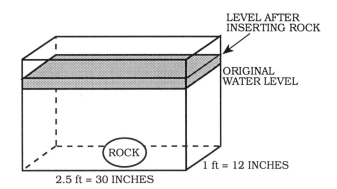

LEVEL AFTER INSERTING ROCK

ORIGINAL WATER LEVEL

ROCK

1 ft = 12 INCHES

2.5 ft = 30 INCHES

Volume of the shaded region = volume of the rock. The volume of the water = $(30)(12)\left(\dfrac{1}{4}\right)$ = 90 cubic inches. Therefore, the volume of the rock is 90 cubic inches.

(b) From part (a) we know that a $\dfrac{1}{4}$ - inch height of water in the aquarium has a volume of 90 cubic inches, so a $\dfrac{1}{2}$ - inch height of water will have a volume that is twice as large, or 180 cubic inches. The volume of the 200 marbles = 200(volume of 1 marble). The volume of 1 marble is $\dfrac{4}{3} \pi(0.75)3 \approx 1.77 \text{ cm}^3$. Therefore, the volume of 200 marbles is 200(1.77) $\approx$ 353.4 cm³. Since 2.54 cm = 1 inch, we know that 16.387 cm³ $\approx$ 1 in³, so 353.4 cm³ $\times \dfrac{1 \text{ in}^3}{16.387 \text{ cm}^3} \approx 21.6 \text{ in}^3$. Therefore, the 21.6 cubic inches of volume occupied by the marbles is far less than the 180 cubic inches needed to overflow the tank. The tank will not overflow.

Section 13.4
19. (a) A cube is one example of a square prism. If the length of the side of the cube is s, then the volume is s^3. If all the dimensions are doubled, then the volume becomes $(2s)^3 = 8s^3$. The volume is increased by a factor of 8.

Volume=S³

Volume= (2S)³ = 8S³

(b) Refer to the diagrams in part (a). The surface area for the original cube is 6 (area of one side) = $6s^2$. The surface area for the new cube is 6 (area of one side) = $6(2s)^2 = 6 (4s^2) = 24s^2$. The surface area is increased by a factor of 4.

Section 13.4

20. The arrangement of pipes which fill the pool the fastest will be the one with the largest combined opening (area).
    (i) Three pipes, each with a diameter of 9 cm or a radius of 4.5 cm, have an area of 3(area of one pipe) = $3(\pi r^2) = 3\pi(4.5)^2 = 60.75\pi$ cm².
    (ii) Two pipes, each with a diameter of 12 cm or a radius of 6 cm, have a combined area of 2(area of one pipe) = $2(\pi r^2) = 2\pi(6)^2 = 72\pi$ cm².
    (iii) One pipe, with a diameter of 16 cm or a radius of 8 cm, has an area = area of one pipe = $1(\pi r^2) = 1\pi(8)^2 = 64\pi$ cm².
    The two 12-inch diameter pipes will fill the pool the fastest.

Section 13.4

21. The volume of a cube is s^3, where s = edge length.
    (a) The volume of a cube with edges of length 2 meters is $(2)^3 = 8$ m³.
    (b) If the volume of a cube is 16 m³, then $16 = s^3$, so $s = 2\sqrt[3]{2}$. The length of an edge is $2\sqrt[3]{2}$ m.

Section 13.4

22. (a) The volume of a circular cylinder is $\pi r^2 h$.

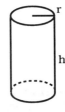

If the radius is doubled and the height remains the same, then the new volume is $\pi(2r)^2 h = \pi \times 4r^2 h = 4\pi r^2 h$. The volume of the new cylinder is increased by a factor of 4.

(b) If the height is doubled and the radius remains the same, then the new volume is $\pi r^2(2h) = 2\pi r^2 h$.. The volume of the new cylinder is doubled.

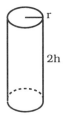

Section 13.4

23. Draw a picture of the prism and let w = width, l = length, and h = height. Now express the areas of the faces in terms of w, l, and h.

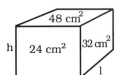

    Area of end: $24 = hw$.
    Area of side: $32 = hl$.
    Area of top: $48 = wl$.

Solve for one variable, and substitute.

Since $24 = hw$, $w = \dfrac{24}{h}$, so $48 = \dfrac{24}{h} \times l$.

Since $32 = hl$, $l = \dfrac{32}{h}$, so $48 = \dfrac{24}{h} \times \dfrac{32}{h}$.

Therefore, $48 = \dfrac{768}{h^2}$.

$$h^2 = \frac{768}{48}$$
$$h^2 = 16$$

$h = 4$, so $w = \dfrac{24}{4} = 6$, and $l = \dfrac{32}{4} = 8$.

The dimensions of the prism are 4 cm by 6 cm by 8 cm. The volume is $4 \times 6 \times 8 = 192$ cm³.

Section 13.4

24. To determine the percentage difference, we need to find the area of the circles with diameters 6 inches (radius 3 inches) and 8 inches (radius 4 inches).

Area of the larger circle = $\pi(4)^2 = 16\pi$ in².

Area of smaller circle = $\pi(3)^2 = 9\pi$ in².

$$\text{Percent difference} = \frac{\text{difference in area}}{\text{smaller area}} \times 100\%$$

$$= \frac{16\pi - 9\pi}{9\pi} \times 100\%$$

$$= \frac{7\pi}{9\pi} \times 100\%$$

$$\approx 78\%.$$

The area of the larger circle is approximately 78% greater than the area of the smaller circle.

Section 13.4

25. The water tank has a diameter of 10 feet (radius 5 feet) and a height of 15 feet. The volume of a cone is $\frac{1}{3}\pi r^2 h$.

The volume of the original tank is $\frac{1}{3}\pi(5)^2(15) = 125\pi$ ft³. The new tank will have the same height but will have half the capacity (volume). The volume of the new tank will be $\frac{1}{2}(125\pi) = 62.5\pi$ ft³. Since $V = \frac{1}{3}\pi r^2 h$, V = 62.5π, and $h = 15$, we can solve for r.

$$62.5\pi = \frac{1}{3}\pi r^2(15)$$

$$r = \sqrt{\frac{3(62.5\pi)}{15\pi}} = \frac{5}{\sqrt{2}} = \frac{5\sqrt{2}}{2}.$$

Therefore, the diameter of the smaller tank is $5\sqrt{2}$ feet ≈ 7.07 feet.

Section 13.4

26. One board foot = (1 ft)(1 ft)(1 inch). Thus, the volume of a board foot is

$$l \times w \times h = (1 \text{ ft}) \times (1 \text{ ft}) \times (\frac{1}{12} \text{ ft})$$

$$= \frac{1}{12} \text{ ft}^3$$

$$= 0.08\overline{3} \text{ ft}^3.$$

Therefore, since 1 board foot = $\frac{1}{12}$ ft³, 1 ft³ = 12 board feet.

(a) A two by four measures $1\frac{1}{2}$" by $3\frac{1}{2}$". If it is 6 feet long (72 inches), then its volume is (1.5)(3.5)(72) = 378 in³, and 378 in³ $\times \frac{1 \text{ ft}^3}{1728 \text{ in}^3}$ = 0.21875 ft³. There are 0.21875 ft³ $\times \frac{12 \text{ board feet}}{1 \text{ ft}^3}$ = 2.625 board feet in this piece of lumber.

(b) A two by eight measures $1\frac{1}{2}$" by $7\frac{1}{2}$". If it is 10 feet long (120 inches), then its volume is $(1.5)(7.5)(120) = 1350$ in³, and $1350 \text{ in}^3 \times \dfrac{1 \text{ ft}^3}{1728 \text{ in}^3} = .78125 \text{ ft}^3$. There are $0.78125 \text{ ft}^3 \times \dfrac{12 \text{ board feet}}{1 \text{ ft}^3} = 9.375$ board feet in this piece of lumber.

(c) Plywood is sold in exact dimensions, so the number of board feet can be calculated directly. A 4-foot by 8-foot by $\frac{3}{4}$ - inch sheet has a volume of $(4)(8)\left(\dfrac{3}{4}\right) = 24$ board feet.

(d) A 4-foot by 6-foot by $\frac{5}{8}$ - inch sheet of plywood has a volume of $(4)(6)\left(\dfrac{5}{8}\right) = 15$ board feet.

Section 13.4

27. We need to calculate the volume of water covering a field 75 m by 135 m to a depth of 3cm (0.03 m). The volume of water is $(75)(135)(0.03) = 303.75$ cubic meters. Since the pump can pump 250 liters of water per minute, we need to find how long it will take to pump 303.75 cubic meters. Convert cubic meters to liters, using dimensional analysis.

$$303.75 \text{ m}^3 \times \frac{1,000,000 \text{ cm}^3}{1 \text{ m}^3} \times \frac{1 \text{ mL}}{1 \text{ cm}^3} \times \frac{1 \text{ L}}{1000 \text{ mL}} = 303750 \text{ L}.$$

A volume of 303.75 cubic meters is the same as 303,750 liters. We can use dimensional analysis to determine the time required by the pump.

$$303,750 \text{ L} \times \frac{1 \text{ minute}}{250 \text{ L}} \times \frac{1 \text{ hour}}{60 \text{ minutes}} = 20.25 \text{ hours}.$$

It will take the pump 20.25 hours.

SOLUTIONS - PART A PROBLEMS

Chapter 14: Geometry Using Triangle Congruence and Similarity

Section 14.1

17. (a) Since, after marking off $\overline{AB}$, Ken walked the same distance to C, $\overline{AB} \cong \overline{AC}$. Both $\angle TAB$ and $\angle DCB$ are right angles and therefore congruent, since they were formed by walking directly to or directly away from the river. From where Ken stands at point D, Betty is lined up with the tree. Therefore, the points D, B, and T are collinear. By vertical angles, $\angle ABT \cong \angle CBD$.

   (b) $\triangle ABT \cong \triangle CBD$ by ASA.

   (c) Since the triangles are congruent and side $\overline{CD}$ of $\triangle CBD$ corresponds to side $\overline{AT}$ of $\triangle ABT$, $\overline{CD} \cong \overline{AT}$. Ken and Betty can find the length of $\overline{CD}$, so they will know the width of the river. (since $\overline{CD} \cong \overline{AT}$)

Section 14.1

18. (a) If you construct similar models, you will find that frameworks (1) and (4) retain their shapes while frameworks (2) and (3) will collapse. Notice how frameworks (2) and (3) can lose their shapes.

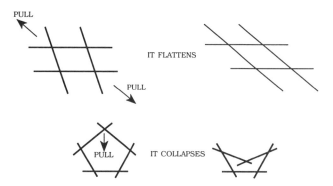

   (b) Because triangles are rigid figures, any scaffolding based on triangles will not collapse.

Section 14.1

19. Every quadrilateral has four sides. Adjacent sides of the quadrilateral meet in a vertex. Since there are no inner supports to keep the distances the same between opposite vertices, every quadrilateral will collapse. Therefore, there are no rigid quadrilaterals.

Section 14.1

20. (a) We can draw two noncongruent triangles that have three pairs of corresponding parts congruent. Consider two different equilateral triangles.

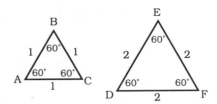

All pairs of corresponding angles of the two triangles are congruent, but the triangles are not congruent.

(b) It is impossible to draw two noncongruent triangles that have four pairs of corresponding parts congruent. Consider all possible situations:

(i) If three pairs of corresponding sides and one pair of corresponding angles are congruent, then the triangles will be congruent by SSS.

(ii) If two pairs of corresponding sides and two pairs of corresponding angles are congruent, then the triangles will be congruent by ASA since the third pair of angles are forced to also be congruent.

(iii) If one pair of corresponding sides and three pairs of corresponding angles are congruent, then the triangles will be congruent by ASA.

(c) It is impossible to draw two noncongruent triangles with five pairs of corresponding parts congruent. Consider all cases:

(i) If three pairs of corresponding sides and two pairs of corresponding angles are congruent, then the triangles are congruent by SSS or ASA.

(ii) If two pairs of corresponding sides and three pairs of corresponding angles are congruent, then the triangles will be congruent by ASA.

(d) It is impossible to draw two noncongruent triangles with six pairs of corresponding parts congruent. If all three corresponding angles are congruent and all three corresponding sides are congruent, then the triangles are congruent by SSS, ASA, or SAS.

Section 14.1

21. (a) $\angle B$ and $\angle K$ are congruent. We know $m(\angle B) = 180° - m\angle A - m\angle C$, $\angle A \cong \angle J$, and $\angle C \cong \angle L$. By substitution we know that $m\angle B = 180° - m\angle J - m\angle L$. However, we also know that $m\angle K = 180° - m\angle J - m\angle L$, so $\angle B \cong \angle K$.

(b) From part (a), we know all three corresponding angles are congruent. Therefore, since $\overline{AB} \cong \overline{JK}$, we can use the congruence property ASA to show $\triangle ABC \cong \triangle JKL$.

(c) With AAS, two angles and the non included side are congruent. When two pairs of angles are congruent, the third pair of angles is automatically congruent. Therefore, AAS is justified.

Section 14.1

22. (a) $\triangle XYZ \cong \triangle BCD$ by SAS since $\overline{BC} \cong \overline{XY}$, $\angle C \cong \angle Y$, and $\overline{CD} \cong \overline{YZ}$.

(b) We are given that $\angle ABC \cong \angle WXY$. Since $\triangle XYZ \cong \triangle BCD$, $\angle 2 \cong \angle 6$ by corresponding parts.
$m\angle ABC = m\angle 1 + m\angle 2$ and $m\angle WXY = m\angle 5 + m\angle 6$
$m\angle 1 + m\angle 2 = m\angle 5 + m\angle 6$ (since $\angle ABC \cong \angle WXY$)
$m\angle 1 + m\angle 6 = m\angle 5 + m\angle 6$ (by substitution)
$m\angle 1 = m\angle 5$ (by subtraction)
So, $\angle 1 \cong \angle 5$.

(c) Since $\triangle XYZ \cong \triangle BCD$, we know $\overline{BD} \cong \overline{XZ}$. Therefore, $\triangle ABD \cong \triangle WXZ$ by SAS. ($\overline{AB} \cong \overline{WX}$, $\angle 1 \cong \angle 5$, and $\overline{BD} \cong \overline{XZ}$.) Corresponding parts that are congruent are $\overline{AD}$ and $\overline{WZ}$.

(d) Since $\triangle ABD \cong \triangle WXZ$, $\angle 3 \cong \angle 7$ by corresponding parts. Since $\triangle BCD \cong \triangle XYZ$, $\angle 4 \cong \angle 8$ by corresponding parts. Therefore, since $m\angle ADC = m\angle 3 + m\angle 4$ and $m\angle WZY = m\angle 7 + m\angle 8$, by substitution we have $m\angle ADC = m\angle 7 + m\angle 8$. Thus, $\angle ADC \cong \angle WZY$.

(e) All corresponding parts of quadrilaterals ABCD and WXYZ have been shown to be congruent. Therefore, ABCD $\cong$ WXYZ.

Section 14.1

23. Suppose $\angle D$ and $\angle E$ are supplementary and congruent. Let $m\angle D = x$. Then $m\angle E = x$. Since the angles are supplementary, $m\angle D + m\angle E = 180°$. By substitution, $x + x = 180°$, or $2x = 180°$, so $x = 90°$. Therefore, $m\angle D = 90°$ and $m\angle E = 90°$.

Section 14.2

7. Set up a proportion using ratios that compare height to waist measurements. Be sure that all units in the ratio are the same. We will need to convert 5 feet 6 inches to inches: 5'6" = 5(12") + 6" = 60" + 6" = 66".

$$\frac{11.5 \text{ inches (doll height)}}{3 \text{ inches (doll waist)}} = \frac{66 \text{ inches (model height)}}{x \text{ inches (model waist)}}$$
$$11.5\, x = (66)\, (3)$$
$$11.5\, x = 198$$
$$x \approx 17.22.$$

For the "life-size" model, the waist measurement would be about 17 inches, an unrealistic measurement for a woman. That is less than the distance around your textbook.

Section 14.2

8. (a) Set up a proportion using ratios that compare thumb height to distance from the projector.

$$\frac{2 \text{ inches}}{5 \text{ feet}} = \frac{x \text{ inches}}{24 \text{ feet}}$$

$$2(24) = 5x$$
$$48 = 5x$$
$$9.6 = x.$$

Therefore, the thumb appears to be 9.6 inches tall on the screen.

(b) Change the measurement in the proportion from part (a) to reflect the change in the distance from the projector.

$$\frac{2 \text{ inches}}{10 \text{ feet}} = \frac{x \text{ inches}}{24 \text{ feet}}$$

$$2(24) = 10x$$
$$48 = 10x$$
$$4.8 = x.$$

Therefore, the thumb appears to be 4.8 inches tall on the screen.

Section 14.2

9. Consider the diagram shown below.

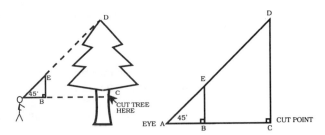

$\triangle ABE$ is a right isosceles triangle. Notice that $m\angle ABE = 90°$, $m\angle EAB = 45°$, and $m\angle BEA = 45°$. Consider $\triangle ACD$. In $\triangle ACD$, $m\angle ACD = 90°$, assuming the tree is vertical and is growing on level ground. We know $m\angle DAC = 45°$ since the top of the tree is exactly in line with the hypotenuse of $\triangle ABE$. $m\angle CDA = 45°$, since the sum of the angles in any triangle is 180°. Therefore, ΔABE ~ ΔACD by the AA Similarity Property. Hence, ΔACD is also a right isosceles triangle. The height, CD, above the cut line is the same length as AC, the distance from where you stand to the tree. The top of the tree will fall where you are standing.

Section 14.2

10. Kathy is 5'6" or 5.5' tall. Notice that ΔABE ~ ΔACD by the AA Similarity Property. We know then that corresponding parts are proportional.

$$\frac{BE}{CD} = \frac{AB}{AC}$$
$$\frac{5.5 \text{ feet}}{30 \text{ feet}} = \frac{x \text{ feet}}{8 \text{ feet}}$$
$$5.5(8) = 30x$$
$$44 = 30x$$
$$1.47 \text{ feet} \approx x.$$

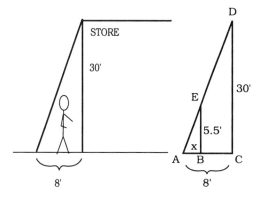

We find that length AB is about 1.47 feet. However, we want to know how far from the building Kathy can stand and still remain in the shade. We need the distance 8 − *x* = 8 − 1.47 = 6.53 feet. Thus, Kathy can stand about 6.53 feet from the building and remain shaded.

Section 14.2

11. (a) Line l_1 contains $\overline{AD}$ and is parallel to line l_2 which contains $\overline{BE}$, so $\overline{AD} \| \overline{BE}$. Transversal *m* contains $\overline{AB}$ and is parallel to transversal *n* which contains $\overline{DE}$, so $\overline{AB} \| \overline{DE}$. Therefore, quadrilateral ABED is a parallelogram because it has two pairs of opposite sides parallel. Similarly, $\overline{BE} \| \overline{CF}$ and $\overline{BC} \| \overline{EF}$, so quadrilateral BCFE is a parallelogram.

(b) Since ABED is a parallelogram such that $\overline{AB}$ and $\overline{DE}$ are opposite sides, we know AB = DE since opposite sides of a parallelogram are congruent. Similarly, since BCFE is a parallelogram, we know that BC = EF.

(c) Complete the proportion by identifying the corresponding parts of the parallelogram. $\overline{AB}$ and $\overline{BC}$ are corresponding parts, and $\overline{DE}$ and $\overline{EF}$ are corresponding parts. Therefore, $\dfrac{AB}{BC} = \dfrac{DE}{EF}$.

(d) Since $\dfrac{AB}{BC} = \dfrac{DE}{EF}$, we see that lines l_1, l_2, and l_3 have intercepted proportional segments on transversals m and n.

Section 14.2

12. (a) Consider the following triangles: $\triangle PAD$, $\triangle PBE$, and $\triangle PCF$. We know $\angle APD \cong \angle BPE \cong \angle CPF$ since the transversals that form the angles for each triangle intersect at P. We also know $\angle PAD \cong \angle PBE \cong \angle PCF$ since these angles are corresponding angles formed when lines l_1, l_2, and l_3, respectively, are cut by transversal m , and we know that corresponding angles are congruent. Therefore, by the AA Similarity Property, we know that $\triangle PAD \sim \triangle PBE \sim \triangle PCF$.

(b) Since corresponding parts of similar triangles are proportional, we know

$$\frac{a}{a+b} = \frac{x}{x+y}$$
$$a\,(x+y) = x\,(a+b)$$
$$ax + ay = ax + bx$$
$$ay = bx \,.$$

(c) Since corresponding parts of similar triangles are proportional, we know

$$\frac{a}{a+b+c} = \frac{x}{x+y+z}$$
$$a\,(x+y+z) = x\,(a+b+c)$$
$$ax + ay + az = ax + bx + cx$$
$$ay + az = bx + cx.$$

(d) From part (b), we know $ay = bx$, and from part (c), we know $ay + az = bx + cx$. Therefore, we can substitute bx for ay.

$$bx + az = bx + cx$$
$$az = cx.$$

(e) From part (b), we know that $ay = bx$. From part (d), we know $az = cx$. Therefore, $\dfrac{bx}{cx} = \dfrac{ay}{az}$ and simplifying gives $\dfrac{b}{c} = \dfrac{y}{z}$.

(f) From part (b), we know $\dfrac{a}{b} = \dfrac{x}{y}$ since $ay = bx$. From part (d), we know $\dfrac{a}{c} = \dfrac{x}{z}$ since $az = cx$. From part (e), we know $\dfrac{b}{c} = \dfrac{y}{z}$. Therefore parallel lines l_1, l_2, and l_3 have intercepted proportional segments on transversals m and n.

Section 14.2

13. (a) Consider parallel lines l_1, l_2, and l_3.

$$\frac{AB}{BC} = \frac{EF}{FG} = \frac{KJ}{JG}.$$

(b) Consider parallel lines l_2, l_3, and l_4.

$$\frac{FG}{GH} = \frac{BC}{CD} = \frac{JG}{GI}.$$

(c) Consider parallel lines l_1, l_2, l_3, and l_4.

$$\frac{IJ}{JK} = \frac{DB}{BA} = \frac{HF}{FE}.$$

Section 14.2

14. We are given that $\overline{DE} \parallel \overline{AB}$, and we can think of segments $\overline{AC}$ and $\overline{BC}$ as transversals. From problem 13, we know parallel lines intercept proportional segments on all transversals. Therefore,

$$\frac{CD}{DA} = \frac{CE}{EB}$$
$$\frac{9}{3} = \frac{CE}{2}$$
$$18 = 3CE$$
$$6 = CE.$$

$CB = CE + EB = 6 + 2 = 8$ and $CA = CD + DA = 9 + 3 = 12$. By the Pythagorean Theorem, $12^2 + 8^2 = (AB)^2$, or $208 = (AB)^2$, so $4\sqrt{13} = AB$.

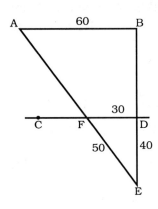

Section 14.2

15. Since $\overline{AB} \| \overline{CD}$ and $\overline{AE}$ and $\overline{EB}$ are transversals, $\angle BAE \cong \angle DFE$, and $\angle ABE \cong \angle FDE$ by corresponding parts. Therefore, $\triangle EBA \sim \triangle EDF$. Since all angles are congruent (Note: $\angle E \cong \angle E$), corresponding parts are proportional.

$$\frac{DE}{FD} = \frac{BE}{AB}$$

$$\frac{40}{30} = \frac{40 + BD}{60}$$

$$\frac{2400}{30} = 40 + BD$$

$$80 - 40 = 40 = BD.$$

Therefore, the canyon is 40 meters wide.

Section 14.2

16. (a) Consider the diagram shown.

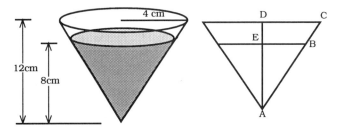

The formula for the volume of a right circular cone is $V = \frac{1}{3}\pi r^2 h$. Consider the cross section of the cone. Since $\triangle ADC \sim \triangle AEB$ we know corresponding parts are proportional.

$$\frac{AD}{AE} = \frac{DC}{EB}$$

$$\frac{12}{8} = \frac{4}{x}$$

$$12x = 32$$

$$x = 2\frac{2}{3} = \frac{8}{3}.$$

Therefore, the radius of the surface of the water is $2\frac{2}{3}$ cm, or the diameter is $5\frac{1}{3}$ cm. The volume of the water is $\frac{1}{3}\pi\left(\frac{8}{3}\right)^2(8) \approx 60\text{cm}^3$.

(b) If the cup is half full of water, then we need to divide the volume of the cup in half.

$$V = \frac{1}{3}\pi(4)^2(12) = 64\pi \text{ cm}^3 \ .$$

One half of the volume is $32\pi \text{ cm}^3$. Find the height when $V = 32\pi$. For a general water level, $\frac{12}{h} = \frac{4}{r}$, so $12r = 4h$ and $r = \frac{1}{3}h$. Now find h when $V = 32\pi$. Since $V = \frac{1}{3}\pi r^2 h$, substitute $V = 32\pi$ and $r = \frac{1}{3}h$.

$$32\pi = \frac{1}{3}\pi\left(\frac{1}{3}h\right)^2 h$$

$$32\pi = \frac{1}{3}\pi\left(\frac{1}{9}h^3\right)$$

$$32\pi = \frac{1}{27}\pi h^3$$

$$32(27) = h^3$$

$$864 = h^3$$

$$9.5 \approx h.$$

Therefore, when the depth of water is about 9.5 cm, the cup will be half full.

Section 14.2

17. Segments $\overline{AB}$ and $\overline{CD}$ intersect at point P. By vertical angles, $\angle APC \cong \angle BPD$. Since $\overline{AB}$ is a vertical segment in the square lattice, $\angle PAC \cong \angle PBD$, and both of them are right angles. Therefore, since two pairs of corresponding angles are congruent, the third pair must also be congruent. Thus, $\triangle PAC \cong \triangle PBD$. Let $AP = x$. Since $AB = 4$, $PB = 4 - x$. We know corresponding sides of similar triangles are proportional. Set up a proportion using the lengths $AC = 1$ and $BD = 4$.

$$\frac{AP}{BP} = \frac{AC}{BD}$$

$$\frac{x}{4-x} = \frac{1}{4}$$

$$4x = 4 - x$$

$$5x = 4$$

$$x = \frac{4}{5} \ .$$

Thus, AP = $\frac{4}{5}$, and BP = $4 - \frac{4}{5} = \frac{16}{5}$. By the Pythagorean

Theorem, $(CP)^2 = 1^2 + \left(\frac{4}{5}\right)^2 = 1 + \frac{16}{25} = \frac{41}{25}$, so $CP = \frac{\sqrt{41}}{5}$.

Also, by the Pythagorean Theorem,

$$(PB)^2 + (BD)^2 = (DP)^2$$
$$\left(\frac{16}{5}\right)^2 + (4)^2 = (DP)^2$$
$$\frac{256}{25} + 16 = (DP)^2$$
$$\frac{256 + 400}{25} = (DP)^2$$
$$\frac{656}{25} = (DP)^2$$
$$4\frac{\sqrt{41}}{5} = DP.$$

Section 14.2
18. Consider a picture.

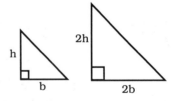

Area of small triangle = $\frac{1}{2}bh$.

Area of large triangle = $\frac{1}{2}(2b)(2h) = \frac{1}{2}(4bh) = 2bh$.

The area of the large triangle is 4 times the area of the small triangle.

Section 14.2
19. (1) $\triangle ACB \sim \triangle CDB$ since corresponding angles are congruent. $m\angle ACB = 90° = m\angle CDB$, and $\angle CBA \cong \angle DBC$, since they are the same angle. Therefore, since two pairs of corresponding angles are congruent, the third pair is also congruent.

(2) Also, $\triangle ACB \sim \triangle ADC$ since corresponding angles are congruent. $m\angle ACB = 90° = m\angle ADC$. $\angle BAC \cong \angle CAD$ since they are the same angle. Therefore, since two pairs of corresponding angles are congruent, the third pair is also congruent.

(3) Therefore, $\triangle CDB \sim \triangle ADC$ since each of them is similar to $\triangle ACB$. Corresponding parts of similar triangles are proportional.

$$\frac{AD}{DC} = \frac{DC}{DB}$$
$$\frac{a}{x} = \frac{x}{1}$$
$$a = x^2$$

So, $\quad \sqrt{a} = x.$

Section 14.2

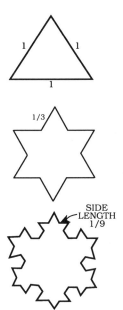

20. (a) The initial Koch curve is a three-pointed star. If each side has length 1, then the perimeter is 3.

(b) The second curve is a six-pointed star. Since each side of the original three-pointed star has the middle third replaced with an equilateral triangle, the side lengths must all be $\frac{1}{3}$. Therefore, the perimeter of the curve is $\frac{1}{3}$ (the number of flat sides) = $\frac{1}{3}(12) = 4$.

(c) On the third curve, each of the flat sides will be $\frac{1}{3}$ as large as in the second curve or have a length of $\frac{1}{3}\left(\frac{1}{3}\right) = \frac{1}{9}$. The perimeter of the third curve is $\frac{1}{9}$ (the number of flat sides) = $\frac{1}{9}(48) = 5\frac{1}{3}$.

(d) To find the perimeter of the nth curve, make a table. In the construction of each curve, every line segment from the previous curve is replaced with four shorter segments. Each of these segments is one-third as long as the line segment from the previous curve.

Curve	Perimeter
1	$3 = 3 \times 1$
2	$4 = 3 \times \left[1 + \dfrac{1}{3}\right]$
3	$5\dfrac{1}{3} = 3 \times \left[1 + \dfrac{1}{3} + 4\left(\dfrac{1}{3^2}\right)\right]$
4	$7\dfrac{1}{9} = 3 \times \left[1 + \dfrac{1}{3} + 4\left(\dfrac{1}{3^2}\right) + 4^2\left(\dfrac{1}{3^3}\right)\right]$
$\vdots$	$\vdots$
n	$3\left[1 + \dfrac{1}{3} + 4\left(\dfrac{1}{3^2}\right) + 4^2\left(\dfrac{1}{3^3}\right) + \ldots + 4^{n-2}\left(\dfrac{1}{3^{n-1}}\right)\right]$

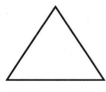

ORIGINAL CURVE

SECOND CURVE

Section 14.2

21. (a) Consider the original curve and the second curve. Notice that the dotted length in the second curve is $\frac{1}{3}$ of the length of the side of the original curve. Recall that the area of a triangle is $\frac{1}{2}$ (base)(height) = $\frac{1}{2}$ *bh*. Notice that the original curve and the small triangle formed by the dotted line in the second curve are both equilateral triangles. Therefore, they are similar, and corresponding parts are proportional. We know that the base of the small triangle is $\frac{1}{3}$ of the base of the original curve. The height of the small triangle is $\frac{1}{3}$ of the height of the original curve. Therefore, the area of the small triangle is

$$\frac{1}{2}\left(\frac{1}{3}b\right)\left(\frac{1}{3}h\right) = \frac{1}{3^2}\left(\frac{1}{2}bh\right) = \frac{1}{9}(\text{area of the original curve}).$$

Since the original curve has an area of 9, each small triangle has an area of 1. Three new small triangles were added to form the second curve, so the area of the second curve is 9 + 3(1) = 12.

(b) For the third curve, the area of each new small triangle is $\frac{1}{9}$ of the area of a small triangle from the second curve, or $\frac{1}{9}$ (1) = $\frac{1}{9}$. Since 12 new triangles were added to form the third curve, the area of the third curve is the area of the second curve plus the areas of the 12 new triangles. Area = $12 + 12\left(\frac{1}{9}\right) = 12 + \frac{4}{3} = 13\frac{1}{3}$.

(c) For the fourth curve, the area of each new small triangle is $\frac{1}{9}$ of the area of a small triangle from the third curve, or $\frac{1}{9}\left(\frac{1}{9}\right) = \frac{1}{81}$. Since there are 4 × 12 = 48 new triangles in the fourth curve, the area of the fourth curve = $13\frac{1}{3} + 48\left(\frac{1}{81}\right) = 13\frac{1}{3} + \frac{16}{27} = 13\frac{25}{27}$.

(d) To find the area of the *n* th curve, make a table. In the construction of each curve, four smaller triangles are added on to each section of the figure. The area of each of the smaller triangles is $\dfrac{1}{3^2}$ or $\dfrac{1}{9}$ as large as the area of a small triangle from the previous curve.

Curve	Area
1	$9 = 3[3]$
2	$12 = 3[3+1]$
3	$13\dfrac{1}{3} = 3\left[3+1+4\left(\dfrac{1}{3^2}\right)\right]$
4	$13\dfrac{25}{27} = 3\left[3+1+4\left(\dfrac{1}{3^2}\right)+4^2\left(\dfrac{1}{3^4}\right)\right]$
5	$14\dfrac{46}{243} = 3\left[3+1+4\left(\dfrac{1}{3^2}\right)+4^2\left(\dfrac{1}{3^4}\right)+4^3\left(\dfrac{1}{3^6}\right)\right]$
.	.
.	.
.	.
n	$3\left[3+1+4\left(\dfrac{1}{3^2}\right)+4^2\left(\dfrac{1}{3^4}\right)+\ldots+4^{(n-2)}\left(\dfrac{1}{3^{2n-4}}\right)\right]$

Section 14.2

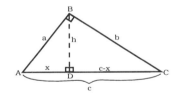

22. Consider the diagram. Notice that there are three right triangles: $\triangle ABC$, $\triangle ADB$, and $\triangle BDC$.

$\angle ABC \cong \angle BDC \cong \angle ADB$, since they are all right angles.

$\angle ACB \cong \angle BCD$, since they are the same angle.

$\angle CAB \cong \angle BAD$, since they are the same angle.

By the AA Similarity Property, we know that $\triangle ABC \sim \triangle BDC$ and $\triangle ABC \sim \triangle ADB$, so all three triangles are similar. Therefore, we know that side lengths are proportional.

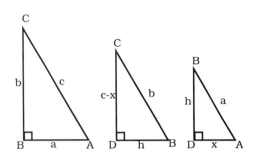

$$\frac{a}{x} = \frac{c}{a} \text{ so } a^2 = cx$$

$$\frac{b}{c-x} = \frac{c}{b} \text{ so } b^2 = c^2 - cx$$

$$a^2 + b^2 = cx + c^2 - cx = c^2$$

Therefore, $a^2 + b^2 = c^2$.

Section 14.3

13. The medians of a triangle extend from each vertex to the middle of the opposite side. Angle bisectors extend from each vertex and bisect the angle. Draw several types of triangles to see when the median and angle bisector coincide.

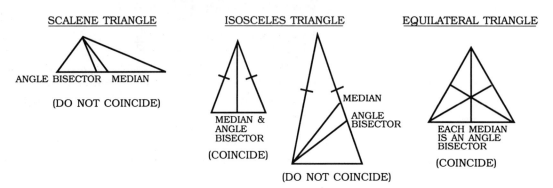

SCALENE TRIANGLE

ANGLE BISECTOR   MEDIAN

(DO NOT COINCIDE)

ISOSCELES TRIANGLE

MEDIAN & ANGLE BISECTOR

(COINCIDE)

MEDIAN

ANGLE BISECTOR

(DO NOT COINCIDE)

EQUILATERAL TRIANGLE

EACH MEDIAN IS AN ANGLE BISECTOR

(COINCIDE)

The median and angle bisector coincide in equilateral triangles and in isosceles triangles for the angle between the congruent sides.

Section 14.3

14. Draw several types of triangles, constructing the medians and the perpendicular bisectors in each of them.

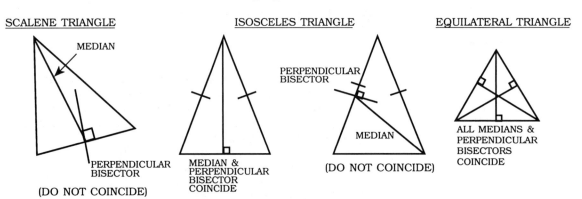

SCALENE TRIANGLE

MEDIAN

PERPENDICULAR BISECTOR

(DO NOT COINCIDE)

ISOSCELES TRIANGLE

MEDIAN & PERPENDICULAR BISECTOR COINCIDE

PERPENDICULAR BISECTOR

MEDIAN

(DO NOT COINCIDE)

EQUILATERAL TRIANGLE

ALL MEDIANS & PERPENDICULAR BISECTORS COINCIDE

The median and perpendicular bisector coincide in equilateral triangles and in isosceles triangles for the angle between the congruent sides.

Section 14.3

15. (a) In the first two steps of the construction, we marked off arcs of radius r from A to B and labeled the points of intersection P and Q. $\triangle APQ$ and $\triangle BPQ$ are formed by drawing in segments. The distances from A to P and B to P are both r. Therefore, $\overline{AP} \cong \overline{BP}$ by

construction. Similarly, $\overline{AQ} \cong \overline{BQ}$ by construction. Since $\overline{PQ} \cong \overline{PQ}$, we know $\triangle APQ \cong \triangle BPQ$ by SSS.

(b) From (a), we know $\overline{AP} \cong \overline{BP}$. Since $\triangle APQ \cong \triangle BPQ$, we know $\angle APR \cong \angle BPR$ by corresponding parts. $\overline{PR} \cong \overline{PR}$ since every segment is congruent to itself. Therefore, $\triangle APR \cong \triangle BPR$ by SAS.

(c) A, R, and B are collinear. $\angle PRB$ and $\angle PRA$ are supplementary so $m\angle PRB + m\angle PRA = 180°$. Since $\triangle APR \cong \triangle BPR$, we know $\angle PRB \cong \angle PRA$ by corresponding parts. Therefore, by substitution, $m\angle PRB + m\angle PRB = 180°$. So $m\angle PRB = 90°$ and $m\angle PRA = 90°$. We conclude $\overline{PR}$ is perpendicular to $\overline{AB}$.

(d) Since $\triangle APR \cong \triangle BPR$, we know corresponding parts are congruent. Therefore, $\overline{AR} \cong \overline{BR}$. Since $\overline{PQ}$ passes through point R, $\overline{PQ}$ bisects $\overline{AB}$.

Section 14.3

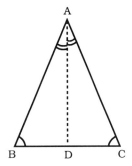

16. Consider $\triangle ABC$ with equal base angles. Construct the bisector of $\angle A$. Label the intersection of the angle bisector and $\overline{BC}$ point D. By the definition of angle bisector, $\angle BAD \cong \angle CAD$. We know $\overline{AD} \cong \overline{AD}$, and $\angle ABD \cong \angle ACD$, so by AAS, $\triangle BAD \cong \triangle CAD$. Therefore, since corresponding parts of congruent triangles are congruent, $\overline{AB} \cong \overline{AC}$, and $\triangle ABC$ is isosceles.

Section 14.4

12. (a) Draw an acute triangle. Recall that the circumcenter of a triangle must be a point on each of the perpendicular bisectors of the sides. Construct each perpendicular bisector, as shown in the next figure. Point D is the circumcenter.

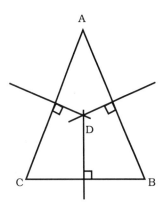

(b)

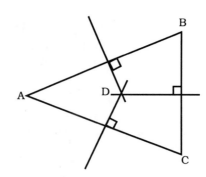

(c) The circumcenter appears to lie inside an acute triangle.

Section 14.4

13. (a) Draw a right triangle. Construct the perpendicular bisectors of each side. Notice that each perpendicular bisector intersects the midpoint of the hypotenuse.

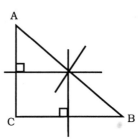

(b)

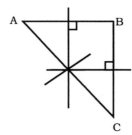

(c) The circumcenter of a right triangle appears to be the midpoint of the hypotenuse.

Section 14.4

14. (a) Draw an obtuse triangle. Construct the perpendicular bisector of each side, as shown in the next figure. Point D is the circumcenter.

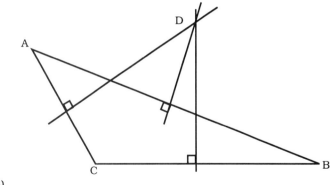

(b)

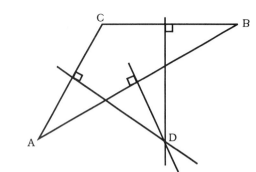

(c) The circumcenter appears to lie outside an obtuse triangle.

Section 14.4

15. Consider the edge of the lined paper. Let it be a transversal. The parallel lines on the page intercept congruent segments on the transversal. Place the plastic strip on the page so that the two corners of one edge lie exactly on lines five apart. This edge of the plastic strip can be thought of as another transversal. Mark five line segments on the plastic strip perpendicular to the length of the strip. Since parallel lines intercept proportional segments on all transversals, the line segments on the plastic strip are proportional to the line segments formed by the parallel lines on the paper. Since the line segments on the plastic strip are proportional to congruent segments, they are congruent.

Section 14.4

16. Consider the construction using a convex quadrilateral and a concave quadrilateral. Notice that the midpoint of a square is the point where the diagonals intersect.

Convex Quadrilateral:

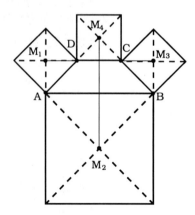

(Note: $\overline{M_1 M_3} \cong \overline{M_2 M_4}$ and $\overline{M_1 M_3} \perp \overline{M_2 M_4}$. All angles formed are right angles.)

Concave Quadrilateral

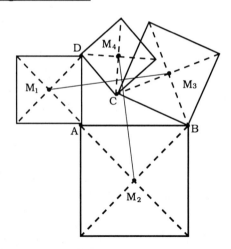

(Note: $\overline{M_1 M_3} \cong \overline{M_2 M_4}$ and $\overline{M_1 M_3} \perp \overline{M_2 M_4}$. All angles formed are right angles.)

(c) $\overline{M_1 M_3} \cong \overline{M_2 M_4}$ and $\overline{M_1 M_3} \perp \overline{M_2 M_4}$.

Section 14.4

17. <u>Case 1:</u> $a > 1$

Let $\overline{AD}$ be a line segment of length a. Let C be any point not on $\overline{AD}$ such that AC = 1. Let E be the point on $\overline{AD}$ such that AE = 1. Connect points C and D. Construct a line, m, through point E so that m is parallel to $\overline{CD}$. Let

B be the point of intersection of m and $\overline{AC}$. We know $\angle AEB \cong \angle ADC$. Since $\angle BAE \cong \angle CAD$, we know that $\triangle ACD \sim \triangle ABE$ by the AA Similarity Property. Let $x = AB$. Using similar triangles $\frac{x}{1} = \frac{1}{a}$, so $x = \frac{1}{a}$.

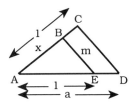

Case 2: $a < 1$

Construct a line segment, $\overline{AE}$, of length 1. Let $\overline{AD}$ be a line segment of length a on the line segment $\overline{AE}$. Let C be any point not on $\overline{AD}$ such that AC = 1. Connect points C and D. Construct a line, m, through point E so that m is parallel to $\overline{CD}$. Let B be the point of intersection of m and $\overleftrightarrow{AC}$. We know that $\angle AEB \cong \angle ADC$. Since $\angle BAE \cong \angle CAD$, we know that $\triangle ACD \sim \triangle ABE$ by the AA Similarity Property. Let $x = AB$. Then, using similar triangles, we have $\frac{x}{1} = \frac{1}{a}$, so $x = \frac{1}{a}$.

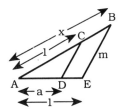

Case 3: $a = 1$

If $a = 1$, then $1 = \frac{1}{a}$, so $\frac{1}{a} = a$ by substitution.

Section 14.4

18. (a) We want to construct the geometric mean of $2 = a$ and $1 = b$. The length 1 is given. Draw a line segment of length greater than 3 and label one endpoint A. Place a compass at A and mark off a length of 1. Label this point B. Place a compass at B and mark off a length of 1. Label this point C. Then we have AC = 2 and BC = 1. Place a compass at B and mark off a length of 2. Label this point D.

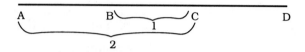

With the compass open to a length of 2, construct arcs from points A and D. Label the intersection of these arcs point E. Construct $\overline{EA}$, $\overline{EB}$, $\overline{EC}$, and $\overline{ED}$. The length of $\overline{EB}$ is x. Since x is the geometric mean of $2 = a$ and $1 = b$ and we know $\dfrac{a}{x} = \dfrac{x}{b}$. If we substitute, then we have $\dfrac{2}{x} = \dfrac{x}{1}$ so $2 = x^2$ and $x = \sqrt{2}$.

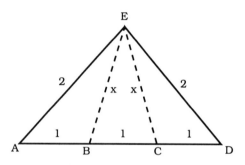

(b) By construction, $\overline{AC} \cong \overline{AE}$ and $\overline{EB} \cong \overline{EC}$, so $\triangle EBC$ and $\triangle AEC$ are isosceles. Therefore, we have that $\angle EBC \cong \angle ECB$ and $\angle ACE \cong \angle AEC$. Now notice that $\angle ECB \cong \angle ACE$ since they are the same angle (just renamed). We see that $\triangle EBC$ and $\triangle AEC$ share the same base angle. Therefore, $\triangle EBC \sim \triangle AEC$ by the AA Similarity Property. Corresponding sides are proportional so $\dfrac{AE}{EB} = \dfrac{EC}{BC}$ or $\dfrac{a}{x} = \dfrac{x}{b}$.

Section 14.5

1. (a) $\angle CAD \cong \angle BAD$, since $\overline{AD}$ bisects $\angle CAB$.

   (b) $\angle B \cong \angle C$, since angles opposite congruent sides of isosceles triangles are congruent.

   (c) ASA

   (d) Corresponding parts of congruent triangles are congruent.

   (e) Since $\angle ADC \cong \angle ADB$ and $\angle ADC$ and $\angle ADB$ are supplementary, the angles are right angles. Therefore, $\overline{AD}$ is perpendicular to $\overline{BC}$.

   (f) Since $\triangle ADC \cong \triangle ADB$, $\overline{CD} \cong \overline{BD}$, since corresponding parts of congruent triangles are congruent. We know

$\overline{AD}$ is perpendicular to $\overline{BC}$. Thus, $\overline{AD}$ is the perpendicular bisector of $\overline{BC}$.

Section 14.5

2. (a) $\overline{AB} \cong \overline{AD}$ and $\overline{BC} \cong \overline{DC}$, since all sides of a rhombus are congruent.
    (b) $\overline{AC} \cong \overline{AC}$, since every segment is congruent to itself.
    (c) SSS
    (d) Since $\triangle ABC \cong \triangle ADC$ by SSS, we know $\angle BAC \cong \angle DAC$ and $\angle BCA \cong \angle DCA$ as they are corresponding parts.
    (e) Since $\angle BAC \cong \angle DAC$, we know $\overline{AC}$ bisects $\angle DAB$.

    Since $\angle BCA \cong \angle DCA$, we know $\overline{AC}$ bisects $\angle DCB$.

Section 14.5

3. $\overline{AB} \cong \overline{CB}$, since all sides of a rhombus are congruent. $\angle ABE \cong \angle CBE$, since vertex angles are bisected by the diagonal. $\overline{BE} \cong \overline{BE}$, since every segment is congruent to itself. Therefore, $\triangle ABE \cong \triangle CBE$ by SAS. $\angle AEB \cong \angle CEB$, since they are corresponding parts of congruent triangles. $m\angle AEB = 90° = m\angle CEB$ as the angles are both supplementary and congruent. Therefore, $\overline{AC}$ is perpendicular to $\overline{DB}$.

Section 14.5

4. Consider lines l, k, m, and n such that $l \,||\, k$ and $m \,||\, n$. Lines m and n are transversals.

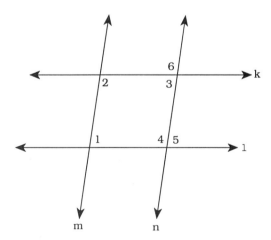

The intersection points of the lines l, k, m, and n are vertices of a parallelogram since the quadrilateral has two sets of opposite sides parallel. $\angle 1 \cong \angle 5$ by corresponding angles. $\angle 4$ and $\angle 5$ are supplementary. Therefore, $\angle 1$ and

∠4 are supplementary, by substitution. Also, ∠4 ≅ ∠6, by corresponding angles, and ∠3 and ∠6 are supplementary. Thus, ∠4 and ∠3 are supplementary, by substitution. Similarly, ∠3 and ∠2 are supplementary, and ∠2 and ∠1 are supplementary. Therefore, consecutive angles of a parallelogram are supplementary.

Section 14.5

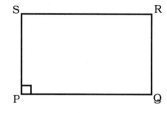

5. We are given parallelogram PQRS with a right angle at P. Since ∠P and ∠Q are consecutive angles in a parallelogram, they are supplementary. $m\angle P + m\angle Q = 180°$. Since ∠P is a right angle, we know that $90° + m\angle Q = 180°$, so $m\angle Q = 90°$. Similarly, ∠Q and ∠R are supplementary, and ∠R and ∠S are supplementary. Therefore, $m\angle R = 90°$ and $m\angle S = 90°$. Parallelogram PQRS has four right angles.

Section 14.5

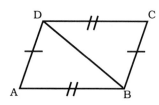

6. Consider quadrilateral ABCD such that $\overline{AD} \cong \overline{BC}$ and $\overline{AB} \cong \overline{DC}$. Construct diagonal $\overline{BD}$. We know $\overline{BD} \cong \overline{BD}$ since every segment is congruent to itself. Therefore, ΔABD ≅ ΔCDB by SSS. Since ΔABD ≅ ΔCDB, we know corresponding parts are congruent, so ∠ABD ≅ ∠CDB and ∠CBD ≅ ∠ADB. Notice ∠ABD and ∠CDB are congruent alternate interior angles formed when lines $\overleftrightarrow{DC}$ and $\overleftrightarrow{AB}$ are cut by transversal $\overleftrightarrow{DB}$. Therefore, $\overleftrightarrow{DC} \parallel \overleftrightarrow{AB}$. Also, ∠CBD and ∠ADB are congruent alternate interior angles formed when lines $\overleftrightarrow{AD}$ and $\overleftrightarrow{BC}$ are cut by transversal $\overleftrightarrow{DB}$. Therefore, $\overleftrightarrow{AD} \parallel \overleftrightarrow{BC}$. Since both pairs of opposite sides of quadrilateral ABCD are parallel, ABCD is a parallelogram.

Section 14.5

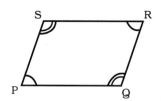

7. Consider quadrilateral PQRS such that ∠P ≅ ∠R and ∠Q ≅ ∠S. We know that the sum of the angles in a quadrilateral is 360°. Thus, $m\angle P + m\angle Q + m\angle R + m\angle S = 360°$.

$$m\angle P + m\angle Q + m\angle P + m\angle Q = 360° \text{ (by substitution).}$$
$$2m\angle P + 2m\angle Q = 360°$$
$$m\angle P + m\angle Q = 180°.$$

Therefore, ∠P and ∠Q are supplementary. Since ∠P and ∠Q are also interior angles on the same side of transversal $\overleftrightarrow{PQ}$, we have $\overleftrightarrow{PS} \parallel \overleftrightarrow{QR}$.

Similarly, $m\angle P + m\angle Q + m\angle R + m\angle S = 360°$
$$m\angle P + m\angle S + m\angle P + m\angle S = 360°$$
$$2m\angle P + 2m\angle S = 360°$$
$$m\angle P + m\angle S = 180°.$$

Therefore, $\angle P$ and $\angle S$ are supplementary.
Since $\angle P$ and $\angle S$ are also interior angles on the same side of transversal $\overleftrightarrow{PS}$, we have $\overleftrightarrow{PQ} \parallel \overleftrightarrow{SR}$. Because both pairs of opposite sides are parallel, PQRS is a parallelogram.

Section 14.5
8. Since quadrilateral ABCD in the text is a parallelogram, we know that $\overline{AB} \parallel \overline{CD}$. Think of $\overline{BD}$ and $\overline{AC}$ as transversals. Because alternate interior angles are congruent, we know that $\angle EBA \cong \angle EDC$ and $\angle EAB \cong \angle ECD$. We also know that AB = CD, since opposite sides of a parallelogram have equal lengths. By the ASA Congruence Property, we know that $\triangle EAB \cong \triangle ECD$. Corresponding parts of congruent triangles are congruent, and, consequently, $\overline{AE} \cong \overline{CE}$ and $\overline{BE} \cong \overline{DE}$. Therefore, the diagonals of a parallelogram bisect each other.

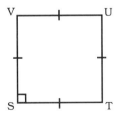

Section 14.5
9. Consider quadrilateral STUV. STUV is a rhombus and $\angle S$ is a right angle. Since every rhombus is a parallelogram, rhombus STUV is a parallelogram with four congruent sides. Since opposite angles in a parallelogram are congruent, $\angle S \cong \angle U$. Therefore, $\angle U$ is a right angle. We also know that adjacent angles are supplementary in a parallelogram. $\angle S$ and $\angle T$ are supplementary.
$$m\angle S + m\angle T = 180°$$
$$90° + m\angle T = 180°$$
$$m\angle T = 90°.$$
Therefore, $\angle T$ is a right angle. Similarly, since $\angle U$ and $\angle V$ are supplementary and $\angle U$ is a right angle, $\angle V$ is a right angle. Quadrilateral STUV has four congruent sides and four right angles. Thus, STUV is a square.

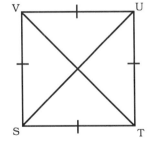

Section 14.5
10. Consider quadrilateral STUV such that STUV is a rhombus and the diagonals are congruent. We know $\overline{ST} \cong \overline{ST}$, $\overline{SV} \cong \overline{TU}$, and we assumed $\overline{SU} \cong \overline{TV}$. By the SSS Congruence Property, $\triangle STV \cong \triangle TSU$. Since corresponding parts of congruent triangles are congruent, $\angle STU \cong \angle TSV$. $\angle STU$ and $\angle TSV$ are also supplementary, since consecutive angles in a parallelogram are supplementary. (Note: Every rhombus is a parallelogram.)

Since ∠STU and ∠TSV are both congruent and supplementary, they are both right angles. Recall that we proved previously that consecutive angles of a parallelogram are supplementary. Therefore, ∠S and ∠V are supplementary, and ∠T and ∠U are supplementary. This forces ∠V and ∠U to both be right angles, since ∠S and ∠T are right angles. Because each angle in STUV is a right angle, and all sides are congruent, STUV is a square.

Section 14.5

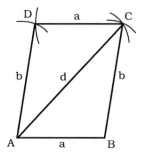

11. Since the diagonal of a parallelogram divides the parallelogram into two congruent triangles, we will begin by constructing one triangle composed of the diagonal and the given side lengths. Construct one side of the triangle using length *a*. Label the endpoints as points A and B. Open the compass to length *d*. Place the compass at A and construct an arc of radius *d*. Open the compass to length *b*. Place the compass at point B and construct an arc of radius *b*. Label the point of intersection of the arcs as point C. Draw ΔABC. Using the SSS Congruence Property, we will complete the parallelogram. Open the compass to length *a*. Place the compass at point C and construct an arc of radius *a*. Open the compass to length *b*. Place the compass at point A and construct an arc of radius *b*. Label the point of intersection of the arcs as D. Draw ΔADC. ABCD is the desired parallelogram.

Section 14.5

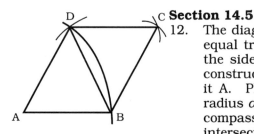

12. The diagonal of a rhombus divides the rhombus into two equal triangles when the diagonal is the same length as the sides. To construct the required rhombus, we will construct two equilateral triangles. Pick a point and label it A. Place the compass at A and construct an arc of radius *a*. Pick a point on the arc and label it B. Place the compass at B and construct an arc of radius *a*. Label the intersection of the arcs D. Construct ΔABD. To complete the rhombus, construct arcs of radius *a* from points B and D. Label the intersection of these arcs as point C. Construct ΔBCD. Quadrilateral ABCD has four sides of length *a*, and diagonal $\overline{BD}$ has length *a*. Therefore, ABCD is the desired rhombus.

Section 14.5

13. Consider an isosceles trapezoid ABCD.

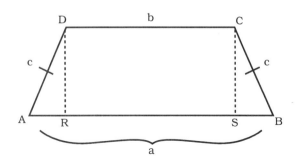

Draw the perpendicular heights from points D and C. Label the points of intersection with the base as R and S. Since AR + RS + SB = a and RS = b, we know AR + SB = $a - b$. Since $\overline{AR} \cong \overline{SB}$, AR = $\dfrac{a-b}{2}$ and SB = $\dfrac{a-b}{2}$. Knowing this, in the construction of trapezoid ABCD we need to find the length $\dfrac{a-b}{2}$ to find points R and S on $\overline{AB}$. At points R and S we will construct perpendicular segments to intersect arcs of radius c from points A and B.

Step 1: Find length $\dfrac{a-b}{2}$.

Construct a segment of length a. Mark off length b.

Bisect the remaining segment. The segment has been divided into three smaller segments of length b, $\dfrac{a-b}{2}$, and $\dfrac{a-b}{2}$.

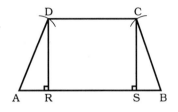

Step 2: Construct the trapezoid.

Construct the base $\overline{AB}$ to have length a. Place the compass at A and construct an arc of radius $\dfrac{a-b}{2}$. Label the intersection on $\overline{AB}$ point R. Place the compass at point B and construct an arc of radius $\dfrac{a-b}{2}$. Label the intersection on $\overline{AB}$ as point S. Construct perpendicular segments at points R and S. Place the compass at point A, and construct an arc of radius c. Label the intersection of the arc and the perpendicular segment as point D. Place the compass at point B and construct an arc of radius c. Label the intersection of the arc and the perpendicular segment as point C. Construct segments $\overline{AD}$, $\overline{DC}$, and $\overline{CB}$. Quadrilateral ABCD is the desired isosceles trapezoid.

Section 14.5

14. We know $\overline{CD}$ is perpendicular to $\overline{AB}$. Since AC is the geometric mean of AD and AB, $\dfrac{AD}{AC} = \dfrac{AC}{AB}$. (In $\triangle ADC$ and $\triangle ACB$, two pairs of corresponding sides are proportional.) $\angle A \cong \angle A$ since every angle is congruent to itself. Therefore, $\triangle ADC \sim \triangle ACB$ by the SAS Similarity Property. Corresponding angles in similar triangles are congruent, so $\angle ADC \cong \angle ACB$. Since $\angle ADC$ is a right angle, $\angle ACB$ is a right angle, and thus $\triangle ABC$ is a right triangle.

Section 14.5

15. (a) Recall that all rectangles have four right angles. This means that adjacent sides are perpendicular. Since each side of the midquad is parallel to a diagonal of the quadrilateral, the diagonals of the quadrilateral must be perpendicular. Therefore, $M_1M_2M_3M_4$ will be a rectangle when diagonals $\overline{PR}$ and $\overline{QS}$ are perpendicular.

    (b) Recall that all four sides of a rhombus have the same length. Because each side of the midquad is half the length of the diagonal, the diagonals must be equal in length. Therefore, $M_1M_2M_3M_4$ will be a rhombus when $\overline{PR} \cong \overline{QS}$.

    (c) To be a square, a quadrilateral must have four right angles, as in part (a), and sides of equal lengths as in part (b). Therefore, $M_1M_2M_3M_4$ will be a square if $\overline{PR}$ and $\overline{QS}$ are perpendicular and congruent.

Section 14.5

16. (a) Since $\angle A \cong \angle A$ and $\angle ADC \cong \angle ACB$, we know $\triangle ADC \sim \triangle ACB$ by the AA Similarity Property.
    (b) Since $\angle B \cong \angle B$ and $\angle BDC \cong \angle BCA$, we know $\triangle BDC \cong \triangle BCA$ by the AA similarity property.
    (c) Since $a^2 = xy + y^2$ and $b^2 = x^2 + xy$,
$$a^2 + b^2 = xy + y^2 + xy + x^2$$
$$= x^2 + 2xy + y^2$$
$$= (x + y)(x + y)$$
$$a^2 + b^2 = (x + y)^2 = c^2.$$
    Therefore, $a^2 + b^2 = c^2$, and the Pythagorean Theorem is proved.

Section 14.5

17. We are given that $\triangle ABC$ and $\triangle A'B'C'$ have three pairs of corresponding sides congruent. However, we cannot use the SSS Congruence Property to conclude that the triangles are congruent. We must show congruence in another way.

    Construct $\triangle ABD$ so that $\angle BAD \cong \angle B'A'C'$ and $AD = A'C'$. We know then that $\triangle BAD \cong \triangle B'A'C'$ by SAS. By congruence of corresponding parts, we know $\overline{BD} \cong \overline{B'C'}$.

    Also, $\overline{BD} \cong \overline{B'C'} \cong \overline{BC}$ by our initial assumption. Now $\triangle ACD$ and $\triangle BCD$ are both isosceles, so their base angles are congruent: $\angle ACD \cong \angle ADC$ and $\angle BCD \cong \angle BDC$. By addition we know $\angle ACB = \angle ACD + \angle BCD$ and $\angle ADB = \angle ADC + \angle BDC$. Substituting, we have $\angle ACB = \angle ADC + \angle BDC$. Therefore, $\angle ADB \cong \angle ACB$ and $\triangle BAD \cong \triangle BAC$ by SAS. Finally, $\triangle B'A'C' \cong \triangle BAC$, since both triangles are congruent to $\triangle BAD$.

Section 14.5

18. We are given that $\angle A \cong \angle A'$. If $\dfrac{AC}{AB} = \dfrac{A'C'}{A'B'}$, then $\triangle ABC \sim \triangle A'B'C'$ by the SAS Similarity Property. If $\dfrac{AC}{AB} \neq \dfrac{A'C'}{A'B'}$, then point D' can be found on $\overline{A'C'}$ such that $\dfrac{AC}{AB} = \dfrac{A'D'}{A'C'}$. Therefore, since $\angle A \cong \angle A'$, we know $\triangle ABC \sim \triangle A'B'D'$ by the SAS Similarity Property. Similar triangles have congruent corresponding angles. Therefore, $\angle ABC \cong \angle A'B'D'$. However, we also know that $\angle ABC \cong \angle A'B'C'$, but $\angle A'B'C'$ and $\angle A'B'D'$ are not the same angle. Since an angle cannot be congruent to two angles of different measures, we have a contradiction. Therefore, $\dfrac{AC}{AB} = \dfrac{A'C'}{A'B'}$ and $\triangle ABC \sim \triangle A'B'C'$ by the SAS Similarity Property.

SOLUTIONS - PART A PROBLEMS

Chapter 15: Geometry Using Coordinates

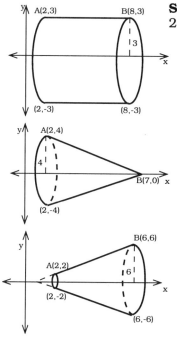

Section 15.1

21. (a) The shape that is generated is a right circular cylinder with radius 3 and height $8 - 2 = 6$. The volume of a cylinder is $\pi r^2 h$.

Volume = $\pi(3)^2(6) = 54\pi$ cubic units.

(b) The shape that is generated is a right circular cone with radius 4 and height $7 - 2 = 5$. The volume of a cone is $\frac{1}{3}\pi r^2 h$.

Volume = $\frac{1}{3}\pi(4)^2(5) = \frac{80}{3}\pi$ cubic units.

(c) The shape that is generated is a right circular cone with the point cut off. The height is $6 - 2 = 4$ and the circular ends have radii 2 and 6 units. The volume of the figure can be found by finding the volume of the complete cone (radius 6, height 6) and subtracting the volume of the (dotted) top (radius 2, height 2).

Volume = $\frac{1}{3}\pi(6)^2 6 - \frac{1}{3}\pi(2)^2 2$

$= 72\pi - \frac{8}{3}\pi$

$= \frac{208\pi}{3}$ cubic units.

Section 15.1

22. (a) Point (2, 1, 3): Move forward 2 units along the x-axis. Move right 1 unit parallel to the y-axis. Move up 3 units parallel to the z-axis.

(b) Point (–2, 1, 0): Move backward 2 units along the x-axis. Move right 1 unit parallel to the y-axis.

(c) Point (3, –1, –2): Move forward 3 units along the x-axis. Move left 1 unit parallel to the y-axis. Move down 2 units parallel to the z -axis.

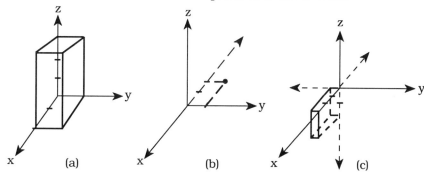

Section 15.1

23. Label the ends of each axis as positive or negative.

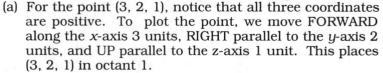

(a) For the point (3, 2, 1), notice that all three coordinates are positive. To plot the point, we move FORWARD along the x-axis 3 units, RIGHT parallel to the y-axis 2 units, and UP parallel to the z-axis 1 unit. This places (3, 2, 1) in octant 1.

(b) To plot (–3, –2, 1), move BACK 3 units, LEFT 2 units, and UP 1 unit. This places (–3, –2, 1) in octant 3.

(c) To plot point (–1, 2, –3), move BACK 1 unit, RIGHT 2 units, and DOWN 3 units. This places (–1, 2, –3) in octant 6.

(d) To plot point (–5, –3, –2), move BACK 5 units, LEFT 3 units, and DOWN 2 units. This places (–5, –3, –2) in octant 7.

(e) To plot point (6, –3, 5), move FORWARD 6 units, LEFT 3 units, and UP 5 units. This places (6, –3, 5) in octant 4.

(f) To plot the point (8, 4, –2), move FORWARD 8 units, RIGHT 4 units, and DOWN 2 units. This places (8, 4, –2) in octant 5.

Section 15.1

24. Label the end of each axis positive or negative.

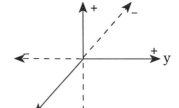

Octant 1: (+, +, +)
Octant 2: (– , +, +)
Octant 3: (– , –, +)
Octant 4: (+, –, +)
Octant 5: (+, +, –)
Octant 6: (–, +, –)
Octant 7: (–, –, –)
Octant 8: (+, –, –)

Section 15.1

25. (a) Each face of the rectangular prism is parallel to one of the coordinate planes. Point R has the same x coordinate as point P since they both lie on a face that is perpendicular to the x-axis. Point R has the same y- and z-coordinates as point Q since they both lie on one face that is perpendicular to the y-axis and one face that is perpendicular to the z-axis. Therefore, point R has coordinates (a, y, z).

(b) The x-coordinate for P and R is the same. However P and R are different points since their y- and z-coordinates are different. We can use the two-dimensional distance formula involving the y- and z-coordinates to find the distance PR.

$$PR = \sqrt{(y-b)^2 + (z-c)^2}$$

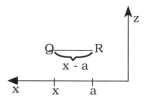

(c) The y- and z-coordinates are the same for points Q and R. Because the points differ only in their x-coordinates, the distance QR is $x - a$.

(d) Notice that $\angle PRQ$ is a right angle, since $\overline{PR}$ and $\overline{QR}$ are on perpendicular faces of the prism. $\overline{PQ}$ is the hypotenuse of the right triangle. Therefore, we can use the Pythagorean Theorem to find the distance PQ.

$$PQ = \sqrt{QR^2 + PR^2}$$
$$PQ = \sqrt{(x-a)^2 + (y-b)^2 + (z-c)^2}$$

Section 15.1

26. If P, M, and Q are collinear with M between P and Q, then PM + MQ = PQ. Calculate the distances using coordinates. Use the coordinates given below.

$$P\,(a,\ b),\quad M\!\left(\frac{a+c}{2},\frac{b+d}{2}\right),\quad \text{and}\quad Q(c,\ d).$$

$$
\begin{aligned}
PM\ &=\ \sqrt{\left(\frac{a+c}{2}-a\right)^2+\left(\frac{b+d}{2}-b\right)^2}\\[2mm]
&=\ \sqrt{\left(\frac{a+c-2a}{2}\right)^2+\left(\frac{b+d-2b}{2}\right)^2}\\[2mm]
&=\ \sqrt{\left(\frac{c-a}{2}\right)^2+\left(\frac{d-b}{2}\right)^2}.
\end{aligned}
$$

$$
\begin{aligned}
MQ\ &=\ \sqrt{\left(c-\frac{a+c}{2}\right)^2+\left(d-\frac{b+d}{2}\right)^2}\\[2mm]
&=\ \sqrt{\left(\frac{2c-a-c}{2}\right)^2+\left(\frac{2d-b-d}{2}\right)^2}\\[2mm]
&=\ \sqrt{\left(\frac{c-a}{2}\right)^2+\left(\frac{d-b}{2}\right)^2}.
\end{aligned}
$$

$$
\begin{aligned}
PM + MQ\ &=\ \sqrt{\left(\frac{c-a}{2}\right)^2+\left(\frac{d-b}{2}\right)^2}\\[2mm]
&\quad +\ \sqrt{\left(\frac{c-a}{2}\right)^2+\left(\frac{d-b}{2}\right)^2}\\[2mm]
&=\ 2\sqrt{\frac{(c-a)^2}{4}+\frac{(d-b)^2}{4}}
\end{aligned}
$$

$$= 2\sqrt{\frac{(c-a)^2 + (d-b)^2}{4}}$$

$$= \frac{2\sqrt{(c-a)^2 + (d-b)^2}}{2}$$

$$= \sqrt{(c-a)^2 + (d-b)^2}$$

Because PQ $= \sqrt{(c-a)^2 + (d-b)^2}$, we see that PM + MQ = PQ, so P, M, and Q are collinear. Also, PM = MQ, since they both equal $\sqrt{\left(\dfrac{c-a}{2}\right)^2 + \left(\dfrac{d-b}{2}\right)^2}$.

Section 15.1

27. Suppose that point (x, y) satisfies the equation $y = mx + b$. Assume (x, y) does not lie on line l. We know $(0, b)$ lies on the line $y = mx + b$ since b is the y-intercept. The slope of the line from the point (x, y) to the point $(0, b)$ is $\dfrac{y-b}{x-0} = \dfrac{y-b}{x}$. However, since (x, y) satisfies the equation $y = mx + b$, if we solve the equation for the slope m, we find $m = \dfrac{y-b}{x}$. Thus, we have two *different* lines that pass through $(0, b)$ and have the same slope [the line l and the line that passed through (x, y)], which is impossible. Therefore, the assumption is false, and (x, y) must be on line l.

Section 15.1

28. Consider distinct lines k and l. Suppose k and l have equal slopes and k and l intersect. Let m be the slope of both k and l. If lines k and l intersect, then they have a point P in common. Let Q be a point on line k and R be a point on line l such that $P \neq Q \neq R$. The slope of $\overline{QP}$ = m, since both points lie on line k. The slope of $\overline{PR}$ = m, since both points lie on line l. Therefore, Q, P, and R must be collinear. However, in order for Q, P, and R to be collinear, the lines l and k must coincide. This contradicts our assumption that the lines are distinct. Consequently, the lines must be parallel.

Section 15.2

32. Sketch the x- and y-axes. Begin measuring the angle at the positive x-axis. Recall how the quadrants are numbered.

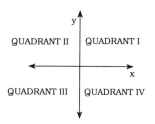

(a) Rotate 45° counterclockwise and plot the point 5 units along the terminal side of the angle.

(b) Rotate 125° counterclockwise and plot the point 3 units along the terminal side of the angle.

(c) Rotate 170° clockwise and plot the point 1 unit along the terminal side of the angle.

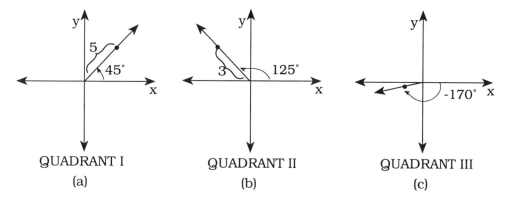

(d) Rotate 240° counterclockwise and plot the point 2 units along the terminal side of the angle.

(e) Rotate 300° counterclockwise and plot the point 1.5 units along the terminal side of the angle.

(f) Rotate 270° clockwise and plot the point 4 units along the terminal side of the angle.

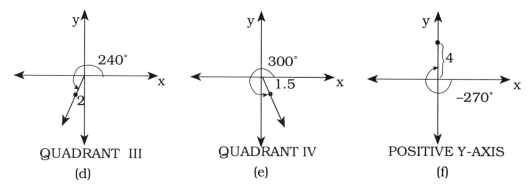

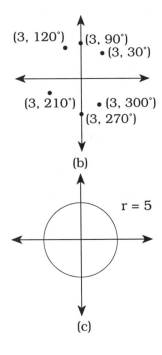

(b)

r = 5

(c)

Section 15.2

33. (a) In the polar coordinate system, a point is identified by (r, θ). To satisfy the equation $r = 3$, the point must be 3 units away from the origin. (Note: No angle is specified.) Any point satisfying $r = 3$ will have 3 as its first coordinate. The following points satisfy the equation: (3, 30°), (3, 90°), (3, 120°), (3, 210°), (3, 270°), and (3, 300°).

(b) Plot the points that satisfy $r = 3$. Notice that the angle can be anything, not just the one of the six angles found in part (a). Each point plotted must be 3 units from the origin. Continue plotting points so that the points are 3 units from the origin but let the angle be any angle. You will notice that the points form a circle of radius 3 centered at the origin.

(c) Points that satisfy the equation $r = 5$ must be 5 units from the origin. The angle can be any angle. The set of points will form a circle of radius 5 centered at the origin.

Section 15.2

34. The point P has coordinates (x, y). From the coordinates given, we can find the lengths of the legs of each triangle formed (see diagram).

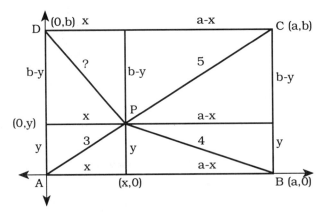

By the Pythagorean Theorem, we can set up equations for appropriate triangles.

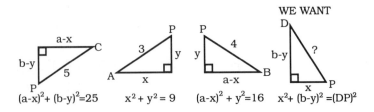

It is impossible to solve for each variable, because we have five unknowns and only four equations. We will try to find a combination of the first three equations to obtain the equation we want. In the equation we want, there is a $(b-y)^2$ term. Begin with the equation

$$(a-x)^2 + (b-y)^2 = 25 \text{ and subtract the equation}$$

$$(a-x)^2 + y^2 = 16 \text{ (to eliminate the } (a-x)^2 \text{ term).}$$

$$(a-x)^2 + (b-y)^2 = 25$$
$$\underline{-(a-x)^2 - y^2 = -16}$$
$$-y^2 + (b-y)^2 = 9$$

Now add the equation $x^2 + y^2 = 9$ to eliminate the y^2 term.

$$-y^2 + (b-y)^2 = 9$$
$$\underline{x^2 + y^2 = 9}$$
$$x^2 + (b-y)^2 = 18$$

Thus, since $x^2 + (b-y) = (DP)^2$, and $x^2 + (b-y)^2 = 18$, we know that $(DP)^2 = 18$, so $DP = 3\sqrt{2}$.

Section 15.2
35. (a) To find the total cost, multiply $2.50 by the number of people and then add the fixed cost of $100.00.

Number of People	30	50	75	100	n
Total cost, y	175	225	287.5	350	$2.5n+100$

  (b) If x = the number of people and y = the total cost, then the linear equation is $y = 2.5x + 100$.
  (c) The slope of the line is 2.5 since the equation is in slope-intercept form. ($y = mx + b$, where m = slope.) The slope represents the cost per person since for every additional person, the total cost increases by $2.50.
  (d) Once again, in the slope-intercept form of the line, $y = mx + b$, where b is the y-intercept. Therefore 100 is the y-intercept of the line. It represents the fixed cost. It must be paid no matter how many people there are.

Section 15.2
36. (a) By the Solutions of Simultaneous Equations Theorem, the equations will have infinitely many solutions if the lines are coincident. This will occur when the slopes and y-intercepts are the same. Write each equation in slope-intercept form.

$$ax + by = c \quad \Rightarrow \quad y = \frac{-ax}{b} + \frac{c}{b}.$$
$$dx + ey = f \quad \Rightarrow \quad y = \frac{-dx}{e} + \frac{f}{e}.$$

If the slopes are the same, then
$$\frac{-a}{b} = \frac{-d}{e} \quad \text{or} \quad ae = bd.$$

If the y-intercepts are the same, then
$$\frac{c}{b} = \frac{f}{e} \quad \text{or} \quad ce = fb.$$

(b) There will be no solution if the lines are parallel, but noncoincident. This will occur when the slopes are the same but the y-intercepts are *not* the same. Consider the slope-intercept form of the lines in part (a). If the slopes are the same, then $ac = bd$. If the y-intercepts are not the same, then $ce \neq fb$.

(c) There will be one solution if the lines intersect in exactly one point. This will occur if the lines have different slopes. If the slopes are not the same, then $ae \neq bd$.

Section 15.3
8. If the diagonals of the rectangle are congruent, then the distances from A to C and B to D are the same. Use the coordinate distance formula to check.

$$\text{AC} = \sqrt{(a - 0)^2 + (b - 0)^2} = \sqrt{a^2 + b^2}$$
$$\text{BD} = \sqrt{(a - 0)^2 + (0 - b)^2} = \sqrt{a^2 + b^2}$$

Since AC = BD, the diagonals are congruent.

Section 15.3
9. Recall that the median of a triangle is the line segment joining a vertex to the midpoint of the opposite side. Consider a diagram of $\triangle ABC$.

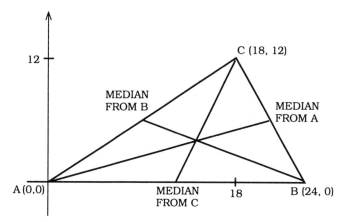

(a) To find the equation of the line containing the median from vertex A, we need the coordinates of the midpoint of $\overline{CB}$.

$$\text{Midpoint of } \overline{CB} = \left(\frac{18+24}{2}, \frac{12+0}{2} \right) = (21, 6)$$

The line will contain the midpoint (21, 6) and the vertex (0, 0). Use the two-point equation of a line.

$$y - y_1 = \frac{y_2 - y_1}{x_2 - x_1}(x - x_1)$$
$$y - 0 = \frac{6 - 0}{21 - 0}(x - 0)$$
$$y = \frac{6}{21}x.$$

The equation of the line is $y = \dfrac{2}{7}x$. We will follow a similar procedure for parts (b) and (c).

(b) Midpoint of $\overline{AC} = \left(\dfrac{0+18}{2}, \dfrac{0+12}{2} \right) = (9, 6).$

The line will contain (9, 6) and the vertex (24, 0).

$$y - 0 = \frac{6 - 0}{9 - 24}(x - 24)$$
$$y = \frac{6}{-15}(x - 24).$$

Thus, $y = \dfrac{-2}{5}x + \dfrac{48}{5}$ is the equation of the line.

(c) Midpoint of $\overline{AB} = \left(\dfrac{0+24}{2}, \dfrac{0+0}{2} \right) = (12, 0)$

The line will contain (12, 0) and the vertex (18, 12).

$$y - 0 = \frac{0-12}{12-18}(x-12)$$
$$y = \frac{-12}{-6}(x-12)$$
$$y = 2(x-12).$$

Thus, $y = 2x - 24$ is the equation of the line.

(d) The intersection of the lines in parts (a) and (b) can be found by substitution.

$$y = \frac{2}{7}x \quad \text{and} \quad y = \frac{-2}{5}x + \frac{48}{5}.$$
$$\frac{2}{7}x = -\frac{2}{5}x + \frac{48}{5}$$
$$10x = -14x + 336$$
$$24x = 336$$
$$x = 14.$$
$$\text{So,} \quad y = \frac{2(14)}{7} = 4.$$

Therefore, the intersection point is (14, 4). If this intersection point lies on the line from part (c), then (14, 4) must be a solution to $y = 2x - 24$. Substitute (14, 4) into the equation to see if the equality holds.

$$4 \stackrel{?}{=} 2(14) - 24$$
$$4 \stackrel{?}{=} 28 - 24$$
$$4 = 4$$

Thus, the point (14, 4) lies on the line $y = 2x - 24$.

(e) The medians all intersect at the same point. Therefore, they are concurrent.

Section 15.3

10. M is the center of the triangle from problem 9. M is the point (14, 4)

(a) Using the distance formula and information from problem 9, we can find the length of the median from A. We need to find the distance between the points A(0, 0) and (21, 6). The length of the median from A is

$$\sqrt{(21-0)^2 + (6-0)^2} = \sqrt{441 + 36} = 3\sqrt{53}.$$

The length of $\overline{AM}$ is the distance from (0, 0) to (14, 4).
$$\sqrt{(14-0)^2 + (4-0)^2} = \sqrt{196 + 16} = \sqrt{212} = 2\sqrt{53}.$$

Therefore, the ratio of the distances $= \frac{2\sqrt{53}}{3\sqrt{53}} = \frac{2}{3}.$

(b) The length of the median from B involves the distance from (24, 0) to (9, 6).

$$\sqrt{(9-24)^2 + (6-0)^2} = \sqrt{225+36} = \sqrt{261} = 3\sqrt{29}.$$

The length of $\overline{BM}$ is the distance from (24, 0) to (14, 4).

$$\sqrt{(14-24)^2 + (4-0)^2} = \sqrt{100+16} = \sqrt{116} = 2\sqrt{29}.$$

The ratio of these distances = $\dfrac{2\sqrt{29}}{3\sqrt{29}} = \dfrac{2}{3}.$

(c) The length of the median from C involves the distance from (18, 12) to (12, 0).

$$\sqrt{(12-18)^2 + (0-12)^2} = \sqrt{36+144} = \sqrt{180} = 6\sqrt{5}.$$

The length of $\overline{CM}$ is the distance from (18, 12) to (14, 4).

$$\sqrt{(14-18)^2 + (4-12)^2} = \sqrt{16+64} = \sqrt{80} = 4\sqrt{5}.$$

The ratio of these distances = $\dfrac{4\sqrt{5}}{6\sqrt{5}} = \dfrac{2}{3}.$

(d) Along any median, the centroid is two-thirds of the distance from the vertex point to the midpoint of the opposite side.

Section 15.3

11. (a) Since AB = BC, we know that the distance from point A(0, 0) to B(a, b) is equal to the distance from B(a, b) to C(c, 0) .

$$AB = \sqrt{(a-0)^2 + (b-0)^2} = \sqrt{a^2 + b^2}.$$
$$BC = \sqrt{(c-a)^2 + (0-b)^2}$$
$$= \sqrt{(c-a)^2 + b^2}.$$

Since AB = BC, we have

$\sqrt{a^2 + b^2} = \sqrt{(c-a)^2 + b^2}.$ Now square both sides.

$a^2 + b^2 = (c-a)^2 + b^2$ Now subtract b^2.

$a^2 = (c-a)^2$ Now multiply.

$a^2 = c^2 - 2ac + a^2$ Now subtract a^2.

$2ac = c^2$

$2a = c.$

(b) The median from B intersects the midpoint of $\overline{AC}$.

The midpoint of $\overline{AC} = \left(\dfrac{0+c}{2}, \dfrac{0+0}{2}\right) = \left(\dfrac{c}{2}, 0\right)$. Since c

$= 2a$, the midpoint is $\left(\dfrac{2a}{2}, 0\right)$ or $(a, 0)$. Therefore, the median from B is vertical, since the x coordinates are the same. The vertical median is perpendicular to the horizontal segment $\overline{AC}$. The line containing the altitude is $x = 8$.

Section 15.3

12. Draw $\triangle RST$. To find the equation of the line containing the altitude, we need to find the line through the vertex, R, and perpendicular to the opposite side. To accomplish this, we need the slope of the opposite side, and we will use the fact that the product of the slopes of perpendicular lines is -1.

(a) The side opposite vertex R is $\overline{ST}$. The slope of $\overline{ST}$ is

$$\dfrac{y_2 - y_1}{x_2 - x_1} = \dfrac{0 - 6}{11 - 8} = -\dfrac{6}{3} = -2.$$ Therefore, a line perpendicular to $\overline{ST}$ will have slope $\dfrac{1}{2}$ (since $(-2)\left(\dfrac{1}{2}\right) = -1$). Thus, the equation of the line through R(0, 0) with slope $\dfrac{1}{2}$ is $y - 0 = \dfrac{1}{2}(x - 0)$, or $y = \dfrac{1}{2}x$.

(b) The opposite side opposite vertex S is $\overline{RT}$. The slope of $\overline{RT}$ is $\dfrac{0 - 0}{0 - 11} = 0$. A line with slope 0 is a horizontal line. A line perpendicular to $\overline{RT}$ must be vertical. Recall that all vertical lines are of the form $x = n$, where n is a constant. The vertical line must pass through S(8, 6), which has x-coordinate 8. Therefore, the line containing the altitude is $x = 8$.

(c) The side opposite vertex T is $\overline{RS}$. The slope of $\overline{RS}$ is $\dfrac{6 - 0}{8 - 0} = \dfrac{6}{8} = \dfrac{3}{4}$. A line perpendicular to $\overline{RS}$ will have slope $-\dfrac{4}{3}$ (since $\left(\dfrac{3}{4}\right)\left(\dfrac{-4}{3}\right) = -1$). Therefore, the equation of the line through T(11, 0) that has slope $\dfrac{-4}{3}$ will be

$$y - 0 = \dfrac{-4}{3}(x - 11) \text{ or } y = \dfrac{-4}{3}x + \dfrac{44}{3}.$$

(d) To find the intersection of the lines $y = \frac{1}{2}x$ and $x = 8$, substitute $y = \frac{1}{2}(8) = 4$. Therefore, the point of intersection occurs at (8, 4). If (8, 4) lies on the line $y = \frac{-4}{3}x + \frac{44}{3}$, then substituting 8 for x and 4 for y will yield a true statement.

$$4 \stackrel{?}{=} \frac{-4}{3}(8) + \frac{44}{3}$$
$$4 \stackrel{?}{=} \frac{-32}{3} + \frac{44}{3}$$
$$4 \stackrel{?}{=} \frac{12}{3}$$
$$4 = 4$$

Therefore, (8, 4) lies on the line. This point is called the orthocenter.

Section 15.3

13. Refer to the picture in the textbook. Label the intersection of the diagonals as point E. If the diagonals of the parallelogram bisect each other, then point E will be the midpoint of $\overline{AC}$ *and* the midpoint of $\overline{BD}$. Find the midpoints of both of these segments.

$$\text{Midpoint of } \overline{AC} = \left(\frac{0 + a + b}{2}, \frac{0 + c}{2}\right) = \left(\frac{a + b}{2}, \frac{c}{2}\right).$$
$$\text{Midpoint of } \overline{BD} = \left(\frac{b + a}{2}, \frac{c + 0}{2}\right) = \left(\frac{a + b}{2}, \frac{c}{2}\right).$$

Notice that the midpoint of $\overline{AC}$ is the same as the midpoint of $\overline{BD}$. We can conclude that point E, the intersection of the diagonals, is the midpoint of both diagonals. Therefore, the diagonals of a parallelogram bisect each other.

Section 15.3

14. (a) If $\overline{MN} \parallel \overline{AC}$, then they will have the same slope. In order to find the slope of $\overline{MN}$, we need to find the coordinates of M and N. Use the fact that they are the midpoints of $\overline{AC}$ and $\overline{BC}$, respectively.

Coordinates of M = $\left(\dfrac{0+a}{2}, \dfrac{0+b}{2}\right) = \left(\dfrac{a}{2}, \dfrac{b}{2}\right)$.

Coordinates of N = $\left(\dfrac{a+c}{2}, \dfrac{b+0}{2}\right) = \left(\dfrac{a+c}{2}, \dfrac{b}{2}\right)$.

Using these coordinates, we can now determine the slope of $\overline{MN}$ and $\overline{AC}$.

Slope of $\overline{MN}$ = $\dfrac{\dfrac{b}{2} - \dfrac{b}{2}}{\dfrac{a+c}{2} - \dfrac{a}{2}} = \dfrac{0}{\dfrac{c}{2}} = 0$.

Slope of $\overline{AC}$ = $\dfrac{0-0}{c-0} = \dfrac{0}{c} = 0$.

Therefore, since $\overline{MN}$ and $\overline{AC}$ both have slopes of zero, we know that they are both horizontal segments and are parallel.

(b) Use the distance formula to find the lengths of $\overline{MN}$ and $\overline{AC}$.

MN = $\sqrt{\left(\dfrac{a+c}{2} - \dfrac{a}{2}\right)^2 + \left(\dfrac{b}{2} - \dfrac{b}{2}\right)^2} = \sqrt{\left(\dfrac{c}{2}\right)^2} = \dfrac{c}{2}$.

AC = $\sqrt{(c-0)^2 + (0-0)^2} = \sqrt{c^2} = c$.

Since AC = c and MN = $\dfrac{c}{2}$, we can see that MN = $\dfrac{1}{2}$AC.

Section 15.3

15. To show that MNOP is a parallelogram, we need to verify that the slopes of opposite sides are the same. To find the slopes, we need the coordinates of M, N, O, and P. Points M, N, O, and P are each midpoints of their respective sides, so we use the midpoint formula to find their coordinates.

M = midpoint of $\overline{DE}$ = $\left(\dfrac{0+a}{2}, \dfrac{0+0}{2}\right) = \left(\dfrac{a}{2}, 0\right)$.

N = midpoint of $\overline{FE}$ = $\left(\dfrac{b+a}{2}, \dfrac{c+0}{2}\right) = \left(\dfrac{b+a}{2}, \dfrac{c}{2}\right)$.

O = midpoint of $\overline{GF}$ = $\left(\dfrac{d+b}{2}, \dfrac{e+c}{2}\right)$.

P = midpoint of $\overline{DG}$ = $\left(\dfrac{0+d}{2}, \dfrac{0+e}{2}\right) = \left(\dfrac{d}{2}, \dfrac{e}{2}\right)$.

Verify that the slopes of $\overline{OP}$ and $\overline{MN}$ are the same and that the slopes of $\overline{PM}$ and $\overline{ON}$ are the same.

The slope of $\overline{OP} = \dfrac{\dfrac{e}{2} - \dfrac{e+c}{2}}{\dfrac{d}{2} - \dfrac{d+b}{2}} = \dfrac{-\dfrac{c}{2}}{-\dfrac{b}{2}} = \dfrac{c}{b}$.

The slope of $\overline{MN} = \dfrac{\dfrac{c}{2} - 0}{\dfrac{b+a}{2} - \dfrac{a}{2}} = \dfrac{\dfrac{c}{2}}{\dfrac{b}{2}} = \dfrac{c}{b}$.

The slopes of $\overline{OP}$ and $\overline{MN}$ are the same, so $\overline{OP} \| \overline{MN}$.

The slope of $\overline{PM} = \dfrac{0 - \dfrac{e}{2}}{\dfrac{a}{2} - \dfrac{d}{2}} = \dfrac{-\dfrac{e}{2}}{\dfrac{a-d}{2}} = \dfrac{-e}{a-d} = \dfrac{e}{d-a}$.

The slope of $\overline{ON} = \dfrac{\dfrac{c}{2} - \dfrac{e+c}{2}}{\dfrac{b+a}{2} - \dfrac{d+b}{2}} = \dfrac{\dfrac{-e}{2}}{\dfrac{a-d}{2}} = \dfrac{-e}{a-d} = \dfrac{e}{d-a}$.

The slopes of $\overline{PM}$ and $\overline{ON}$ are the same, so $\overline{PM} \| \overline{ON}$. Finally, since the quadrilateral has two pairs of opposite sides parallel, MNOP is a parallelogram.

Section 15.3

16. If the diagonals of a square are perpendicular, then the product of the slopes of the line segments will be −1. Find the slopes of each diagonal.

The slope of $\overline{AC} = \dfrac{a-0}{a-0} = \dfrac{a}{a} = 1$.

The slope of $\overline{DB} = \dfrac{0-a}{a-0} = \dfrac{-a}{a} = -1$.

The product of the slopes = $(1)(-1) = -1$. Therefore, the diagonals of the square are perpendicular.

Section 15.3

17. Suppose that $\triangle ABC$ is isosceles. Then suppose $\overline{AC} \cong \overline{BC}$. Show that the medians from A and B are congruent. First find the midpoints of sides $\overline{AC}$ and $\overline{BC}$. Then use the distance formula to show that the medians are the same length.

The midpoint of $\overline{AC} = \left(\dfrac{-a+0}{2}, \dfrac{0+b}{2} \right) = \left(\dfrac{-a}{2}, \dfrac{b}{2} \right).$

The midpoint of $\overline{BC} = \left(\dfrac{a+0}{2}, \dfrac{0+b}{2} \right) = \left(\dfrac{a}{2}, \dfrac{b}{2} \right).$

The distance from A $(-a, 0)$ to the midpoint of $\overline{BC}$ with coordinates $\left(\dfrac{a}{2}, \dfrac{b}{2} \right)$ is

$$\sqrt{\left(\dfrac{a}{2} - (-a) \right)^2 + \left(\dfrac{b}{2} - 0 \right)^2} = \sqrt{\left(\dfrac{3a}{2} \right)^2 + \left(\dfrac{b}{2} \right)^2}.$$

The distance from B$(a, 0)$ to the midpoint of $\overline{AC}$ with coordinates $\left(-\dfrac{a}{2}, \dfrac{b}{2} \right)$ is

$$\sqrt{\left(\dfrac{-a}{2} - a \right)^2 + \left(\dfrac{b}{2} - 0 \right)^2} = \sqrt{\left(\dfrac{3a}{2} \right) + \left(\dfrac{b}{2} \right)^2}.$$

Therefore, since the distances are the same, the medians to the congruent sides are congruent.

Section 15.3

18. (a) Point P has coordinates $(0, 0)$. Since line l is perpendicular to $\overline{QR}$, and we know the coordinates of Q(a, b) and R$(c, 0)$, we find the slope of the line segment $\overline{QR}$ and recall that any segment perpendicular to it will have a slope which is its negative reciprocal.

The slope of $\overline{QR} = \dfrac{0-b}{c-a} = \dfrac{-b}{c-a}$

The slope of $l = \dfrac{c-a}{b}$

(b) To find the intersection of lines l and m, we need to solve the equations simultaneously. The equation of line m is $x = a$, since it is a vertical line passing through Q(a, b). The equation of line l is $y = \dfrac{c-a}{b} x$, since from part (a), we know that the slope is $\dfrac{c-a}{b}$ and the y-intercept is 0. Substituting, we have

$$y = \frac{c-a}{b}(a) = \frac{(c-a)a}{b}$$

where $x = a$, and $y = \frac{(c-a)a}{b}$. Therefore, the lines

l and m intersect at the point $\left(a, \frac{(c-a)a}{b}\right)$.

(c) Since n is perpendicular to $\overline{PQ}$, and we know the coordinates of $P(0, 0)$ and $Q(a, b)$ we can find the slope of

$\overline{PQ} = \frac{b-0}{a-0} = \frac{b}{a}$. The slope of a line segment perpendicular to $\overline{PQ}$ will have slope $\frac{-a}{b}$.

Section 15.3

19. In an equilateral triangle, all three sides are the same length. Consider a square lattice. Construct a triangle and assume it is equilateral. Because the horizontal side has an arbitrary length, we can give it a variable length. Let $2a = $ the length of the horizontal side. Notice $2a$ must be a whole number, since it is the distance between two dots. The other two legs of the triangle must also have length $2a$. Let $b = $ the height of the triangle. Notice that b must also be a whole number, since it is the perpendicular distance from a vertex to the base of the triangle. With the construction of the height, two right triangles are formed, each with height b, base a, and hypotenuse $2a$. By the Pythagorean Theorem,

$$a^2 + b^2 = (2a)^2$$
$$a^2 + b^2 = 4a^2$$
$$b^2 = 3a^2$$
$$\frac{b^2}{a^2} = 3$$
$$\frac{b}{a} = \sqrt{3} \quad \text{(an irrational number).}$$

Since a is a whole number and $\frac{b}{a}$ is irrational, b must be irrational. (If b were rational, then $\frac{b}{a}$ would be rational by the closure property of rational number division.) This contradicts the assumption that b is a whole number. Therefore, b is irrational and will not connect a vertex with the base. We conclude that an equilateral triangle with a horizontal side cannot be formed on a square lattice.

Section 15.3

20. Use a variable to represent each girl's age. Let $x =$ the older girl's age and $y =$ the younger girl's age. Then, since the sum of their ages is 18, $x + y = 18$. Since the difference of their ages is 4, $x - y = 4$, Solve simultaneously by adding the two equations.

$$x + y = 18$$
$$\underline{x - y = 4}$$
$$2x = 22$$
$$x = 11.$$

Since $x + y = 18$ and $x = 11$, $11 + y = 18$, so $y = 7$. Therefore, the ages of the two girls are 7 and 11 years.

Section 15.3

21. Let $x =$ the number of bicycles that pass the house. Let $y =$ the number of tricycles that pass the house. Since every bicycle has two wheels, the number of bicycle wheels that pass by is $2x$. Since every tricycle has three wheels, the number of tricycle wheels that pass by is $3y$. Only seven riders pass by, so $x + y = 7$. There were a total of 19 wheels that passed by, so $2x + 3y = 19$. Solve these two equations by substitution.

$$x + y = 7 \qquad \text{so} \qquad x = 7 - y$$
$$2(7 - y) + 3y = 19$$
$$14 - 2y + 3y = 19$$
$$y = 5.$$

If $y = 5$, then $x = 7 - 5 = 2$. Therefore, there were 2 bicycles and 5 tricycles.

Section 15.3

22. Let $x =$ the amount of money Mike invested, and $y =$ the amount of money Joan invested. Since they invested $11,000 together, $x + y = 11,000$. After Mike triples his investment and Joan doubles hers, their total investment is $29,000, so $3x + 2y = 29,000$. Solve the system by substitution.

$$x + y = 11000 \qquad \text{so} \quad x = 11000 - y$$
$$3(11000 - y) + 2y = 29000$$
$$33000 - 3y + 2y = 29000$$
$$33000 - y = 29000$$
$$4000 = y$$
$$x = 11,000 - 4000 = 7000$$

Therefore, Mike invested $7000, and Joan invested $4000.

Section 15.3

23. Let x = the number of quarters used, and y = the number of dimes used. Since the value of a quarter is 25 cents, the total value of x quarters is $0.25x$ dollars. Likewise, the total value of y dimes is $0.10y$ dollars. Since $16.25 was paid, $0.25x + 0.10y = 16.25$. She used all 110 coins, so $x + y = 110$. Solve the system by substitution.

$$x + y = 110 \qquad \text{so} \qquad x = 110 - y$$
$$0.25\,(110 - y) + 0.10y = 16.25$$
$$27.5 - 0.25y + 0.10y = 16.25$$
$$27.5 - 0.15y = 16.25$$
$$11.25 = 0.15y$$
$$75 = y$$
$$x = 110 - 75 = 35$$

Therefore, she had 35 quarters and 75 dimes.

Section 15.3

24. (a) Consider a simpler problem. Construct a 2×3 rectangular network of paths. Label each vertex. In a table list the length of the path from A to each vertex and the number of paths from A to each vertex.

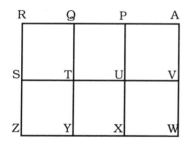

From A to	Length of path	Number of of paths
P	1	1
Q	2	1
R	3	1
S	4	4
T	3	3
U	2	2
V	1	1
W	2	1
X	3	3
Y	4	6
Z	5	10

As an example, the next diagram illustrates all of the paths from A to Y.

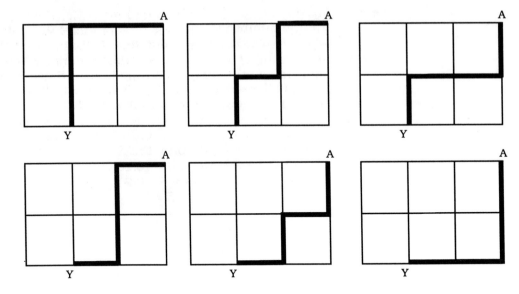

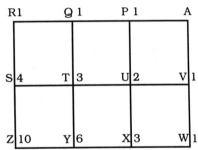

On the 2 × 3 rectangular network of paths, label each vertex with the number of paths you found to it from A. Notice that the number of ways to get from A to each vertex is the sum of the number of ways to get from A to the vertices directly above and to the right. This can be generalized to the case where we have a 4 × 3 rectangular network of paths. Therefore, there are 35 paths of length 7 from A to C.

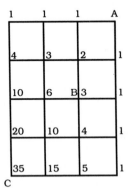

(b) By the Fundamental Counting Principle, the number of paths from A to C through B is the product of the number of paths from A to B and the number of paths from B to C. The number of paths from A to B is 3. To find the number of paths from B to C, draw a 2 × 2 network and label the vertices as we have done previously. There are 6 paths from B to C. Therefore, there are 3 × 6 = 18 paths from A to C through B.

(c) Find the probability that the path will go through B by considering the ratio comparing the number of paths from A to C through B to the total number of paths from A to C. From parts (a) and (b) we know the ratio is

$$\frac{18}{35} \approx 0.51.$$

Section 15.3

25. To demonstrate the placement of various squares, consider a simpler problem. Suppose you have a 3 × 3 square board. There are four 1 × 1 squares and one 2 × 2 square.

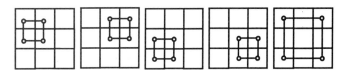

(a) For the 5 × 5 square board, consider the possible square sizes. Make a list of the number of possible locations for one vertex of each size square.

Square Size	Number of Squares
1 × 1	$16 = (5 - 1)^2$
2 × 2	$9 = (5 - 2)^2$
3 × 3	$4 = (5 - 3)^2$
4 × 4	$1 = (5 - 4)^2$

Total: $30 = n(n + 1)$

(b) Consider the upper left side of the tilted square. As the square is "tilted", the slope of the upper left side changes. Count all possible squares (large and small) for a given slope. The possible slopes are shown in the next diagram.

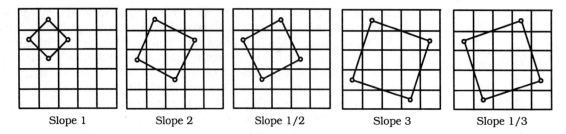

| Slope 1 | Slope 2 | Slope 1/2 | Slope 3 | Slope 1/3 |

Slope	Number of Squares
1	10 (9 small, 1 large)
2	4
1/2	4
3	1
1/3	1
	Total = 20

If we include "tilted" squares, then there are 30 + 20 = 50 different squares which can be formed.

(c) Draw pictures and use what you learned from parts (a) and (b) to help you count all possible squares for a 6 × 6 square.

Number of squares with horizontal or vertical sides

Square Size	Number of Squares
1 × 1	$25 = (6 - 1)^2$
2 × 2	$16 = (6 - 2)^2$
3 × 3	$9 = (6 - 3)^2$
4 × 4	$4 = (6 - 4)^2$
5 × 5	$1 = (6 - 5)^2$
	Total = 55

Number of squares with tilted sides

Slope	Number of Squares
1	20 (16 small, 4 large)
2	9
1/2	9
3	4
1/3	4
4	1
1/4	1
	Total = 48

In a 6 × 6 square, we can form 55 + 48 = 103 squares.

Section 15.3

26. Let x = the number of rows of black circles and y = the number of columns of black circles. The total number of rows of white *and* black circles is $x + 2$, and the total number of columns of white *and* black circles is $y + 2$. Altogether, the total number of circles is $(x + 2)(y + 2)$. The number of white circles = the total number of circles – the number of black circles.

$$\begin{aligned}
\text{Number of white circles} &= (x + 2)(y + 2) - xy \\
&= xy + 2x + 2y + 4 - xy \\
&= 2x + 2y + 4.
\end{aligned}$$

Now, is it possible for the total number of white circles, $(2x + 2y + 4)$, to be the same as the number of black circles, xy? Construct a table and list possible numbers for x and y. Begin by letting $x = 3$. Since we are wondering if it is possible for the number of black circles, xy, to equal the number of white circles, $2x + 2y + 4$, set these expressions equal to each other. Substitute the value for x and solve to find the value of y. Then find xy and $2x + 2y + 4$.

x	y	xy	$2x + 2y + 4$
3	10	30	30
4	6	24	24
5	14/3	(Values in the columns cannot be fractions.)	
6	4	24	24 (same dimensions as above)

Since the number of rows is $x + 2$, and the number of columns is $y + 2$, the mats could have dimensions 5 by 12 or 6 by 8.

SOLUTIONS - PART A PROBLEMS

Chapter 16: Geometry Using Transformations

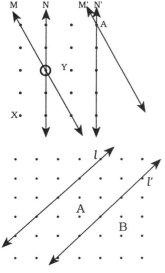

Section 16.1

35. Notice in translation T_{XY}, the point X is moved up 2 units and right 2 units. Therefore, if we want to find a line whose image under T_{XY} passes through point A, we need to find the point that is translated to A under T_{XY}. Move left 2 units and down 2 units to the point in row 3, column 2 (just to the left of Y). Any line which passes through that point will have an image under T_{XY} that passes through point A. Note that there are many solutions.

Section 16.1

36. (a) The translation T_{AB} translates each point down 1 unit and to the right 2 units. Thus, each lattice point on line l will move down 1 unit and right 2 units.
    (b) Yes, the translation image of a line is a line.
    (c) Since all points on the original line undergo the same translation, the line and its translation image are parallel.

Section 16.1

37. (a) Draw a line segment connecting a lattice point on line l to point O. Turn the line segment 90° about point O to find the image of the point. Do this for each lattice point on line l. Notice that the image is also a line. The rotation of two of the lattice points is done below. Since each lattice point is rotated in the same manner, the rotation image of a line is a line.

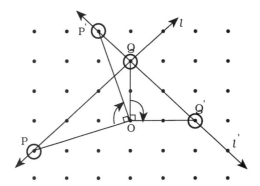

(b) The reflection line is vertical. To find the image of a point under a reflection, draw a line from a lattice point on *l* perpendicular to the reflection line. The image of the point will be on the line an equal distance away from the reflection line but on the other side of the reflection line. Do this to find the image of each lattice point. The image points all lie on line *l'*, so the reflection of a line is also a line. The reflections of two lattice points are shown.

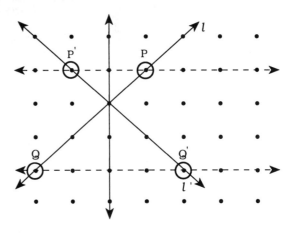

Section 16.1

38. (a) If A' is the image of A under the translation , we must move 2 units down and 2 units right. This translation is denoted by $T_{AA'}$.

(b) Any rotation needs a point about which the points are rotated (the center of rotation) and an angle of rotation. The center must be equidistant from A and A'. The only three such points are along the diagonal of the lattice from the lower left to the upper right. Three rotations that take point A to A' are shown.

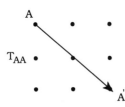

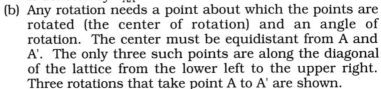

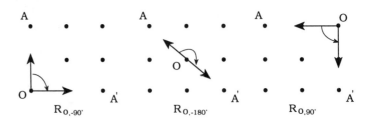

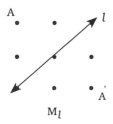

M_l

(c) Connect A and A'. The line of reflection must be perpendicular to this segment joining the point and its image. Since the segment $\overline{AA'}$ is the diagonal of the square lattice, the reflection line is the other diagonal.

(d) The glide reflection is formed by translating A and then reflecting it over a line parallel to the directed line segment of the translation. In this case, the translation line and the reflection line must both be vertical or horizontal.

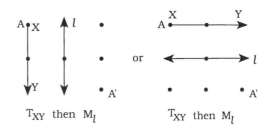

T_{XY} then M_l $\qquad$ T_{XY} then M_l

Section 16.1

39. (a) Since a line is translated parallel line under a translation, there is no translation that maps $\overline{AB}$ to $\overline{A'B'}$. Notice that $\overline{AB}$ and $\overline{A'B'}$ intersect.

(b) Since the center of rotation is equidistant from each point and its image, the center of rotation must be A. (Under the rotation, A cannot move since A and A' are the same point. The rotation must send the rest of the points on $\overline{AB}$ to its image, $\overline{A'B'}$, which is the same as $\overline{AB'}$. Thus, point A is the only choice for the center of rotation.)

(c) The line of reflection must be perpendicular to the line joining any point and its image. Since A and A' are the same point, consider the line joining B and B'. Because it is vertical, the line of reflection must be horizontal. Therefore, the reflection line is a horizontal line passing through A.

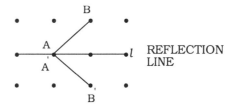

REFLECTION LINE

(d) Recall that each point must be translated and then reflected over a line parallel to the directed line segment of the translation. Notice that since A and A' are the same point, if we translate A in any direction and then reflect it over a parallel line to the direction of translation, we will never transform A to A'. Therefore, there is no glide reflection that maps $\overline{AB}$ to $\overline{A'B'}$.

Section 16.1

40. (a) Draw the coordinate axis. Consider a triangle. Plot the three vertices of the triangle. Transform each vertex to find the image of the original triangle and then determine the reflection line. Keep in mind that the reflection line is perpendicular to the line joining any point and its image and is equidistant from the two of them.

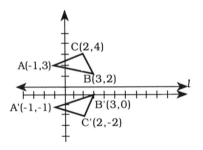

Since $(x, y) \rightarrow (x, 2 - y)$,

$$A(-1, 3) \rightarrow (-1, 2 - 3) = A'(-1, -1)$$
$$B(3, 2) \rightarrow (3, 2 - 2) = B'(3, 0)$$
$$C(2, 4) \rightarrow (2, 2 - 4) = C'(2, -2).$$

Notice the line joining B and B' is vertical since the x coordinate for both is 3. The reflection line l must be horizontal. Since the distance between B and B' is 2 units, the reflection line passes horizontally one unit from each point. Therefore, the equation of the reflection line is $y = 1$.

(b) Consider another triangle.

Since $(x, y) \rightarrow (x, 8 - y)$,

$$A(1, 1) \rightarrow (1, 8 - 1) = A'(1, 7)$$
$$B(-3, 2) \rightarrow (-3, 8 - 2) = B'(-3, 6)$$
$$C(0, -1) \rightarrow (0, 8 - (-1)) = C'(0, 9).$$

Consider points C and C'. The line connecting them is vertical, and the distance between them is $9 - (-1) = 10$. Therefore, the reflection line must be horizontal, passing through the point that is 5 units from each of C and C'. The equation of the reflection line is $y = 4$.

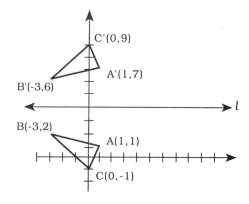

(c) Consider another triangle.

Since $(x, y) \rightarrow (x, -4 - y)$,

$$A(-4, 0) \rightarrow (-4, -4 - 0) = A'(-4, -4)$$
$$B(1, 4) \rightarrow (1, -4 - 4) = B'(1, -8)$$
$$C(4, 2) \rightarrow (4, -4 - 2) = C'(4, -6).$$

Consider points A and A'. The line connecting them is vertical, and the distance between them is 4. Therefore, the reflection line must be horizontal, passing through the point that is 2 units from each of A and A'. The equation of the reflection line is $y = -2$.

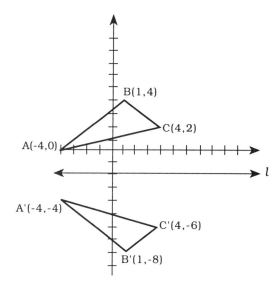

Section 16.1

41. (a) Consider the original points A, B, and C. If we construct horizontal lines through these points, the images of A, B, and C will lie on the opposite side of the reflection line along the horizontal line. Each point and its reflection will be equidistant from the reflection line.

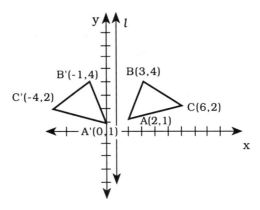

   (b) The coordinates of the image points of A, B, and C under M_l are A' = (0, 1), B'= (-1, 4), and C'(-4, 2).

   (c) Since the reflection line is vertical, the y coordinates will not change when a point is reflected to its image. Consider the x coordinate. Since the point and its image are equidistant from the reflection line, consider the distance from each point to the reflection line. The equation of the reflection line is $x = 1$. Let x represent the x coordinate of any point. The distance from the point to the reflection line is $x - 1$. To find the x coordinate of the image, begin at the original point and subtract the distance $x - 1$ twice.

$$x - 2(x - 1) = -x + 2 = 2 - x.$$

   Therefore, under M_l,

$$(x, y) \to (2 - x, y).$$

Section 16.2

17. (a) We know that $\overrightarrow{PQ}$, $\overrightarrow{BB'}$, $\overrightarrow{XX'}$, and $\overrightarrow{AA'}$ are equivalent directed line segments. Thus, by the definition of the translation T_{PQ}, we have $\overline{PQ} \| \overline{BB'} \| \overline{XX'} \| \overline{AA'}$, and $\overline{PQ} \cong \overline{BB'} \cong \overline{XX'} \cong \overline{AA'}$. Since $\overline{BB'}$ and $\overline{XX'}$ are parallel, congruent, opposite sides of quadrilateral BB'X'X, we know that BB'X'X is a parallelogram. Since $\overline{BB'}$ and $\overline{AA'}$ are parallel, congruent, opposite sides of quadrilateral BB'A'A, we know that the quadrilateral is a parallelogram.

(b) Line $\overleftrightarrow{B'X'}$ ‖ $\overleftrightarrow{BX}$ and $\overleftrightarrow{B'A'}$ ‖ $\overleftrightarrow{BA}$, since these lines contain opposite sides of parallelograms BB'X'X and BB'A'A, respectively. However, since through B' there can be only one line parallel to $\overleftrightarrow{BA}$, we conclude that $\overleftrightarrow{B'X'}$ and $\overleftrightarrow{B'A'}$ must be the same line. So, A', X', and B' are collinear.

Section 16.2

18. Because a translation is an isometry, we know it preserves distances. Therefore, in $\triangle PQR$ and $\triangle P'Q'R'$, PQ = P'Q', QR = Q'R', and RP = R'P'. Since three corresponding sides are the same, by the SSS Triangle Congruence Property, we know that $\triangle PQR \cong \triangle P'Q'R'$. Therefore, since corresponding parts of congruent triangles are congruent, each pair of corresponding angles is congruent. Thus, translations preserve angle measure.

Section 16.2

19. Rotations are isometries, and isometries preserve angle measure. Notice that $\angle 1$ and $\angle 2$ are corresponding angles when parallel lines p and q are cut by transversal m. So, $\angle 1 \cong \angle 2$. Since isometries preserve angle measure, $\angle 1 \cong \angle 3$ and $\angle 2 \cong \angle 4$, so by substitution we know that $\angle 3 \cong \angle 4$. Since $\angle 3$ and $\angle 4$ are congruent corresponding angles, we know p' ‖ q'. Therefore, rotations preserve parallelism.

Section 16.2

20. Since A, B, and C are collinear and B is between A and C, we know that AB + BC = AC. Consider the reflection that maps A to A', B to B', and C to C'. Since reflections are isometries, and isometries preserve distance, we know that AB = A'B', BC = B'C', and AC = A'C'. Therefore, by substitution, we have A'B' + B'C' = A'C', so points A', B', and C' are collinear.

Section 16.2

21. Reflections are isometries and therefore preserve angle measure. If p ‖ q, then corresponding angles are congruent. $\angle 1 \cong \angle 2$ because they are corresponding angles. Since reflections preserve angle measure, we have $\angle 1 \cong \angle 1'$ and $\angle 2 \cong \angle 2'$, so by substitution we know $\angle 1' \cong \angle 2'$. Therefore, since corresponding angles are congruent, we know p' ‖ q'.

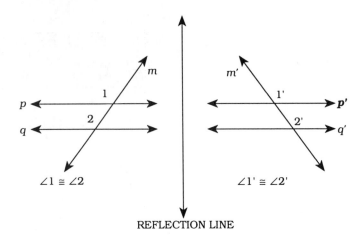

$\angle 1 \cong \angle 2$ $\angle 1' \cong \angle 2'$

REFLECTION LINE

Section 16.2
22. An isometry maps segments to segments and preserves distance. Consider $\triangle ABC$ and $\triangle A'B'C'$. We know that $\overline{AB} \cong \overline{A'B'}$, $\overline{BC} \cong \overline{B'C'}$, and $\overline{CA} \cong \overline{C'A'}$. Since three pairs of corresponding sides are congruent, we know that $\triangle ABC \cong \triangle A'B'C'$ by the SSS Congruence Property.

Section 16.2
23. (a) Since between any two points there is at most one line, there can be only one translation that maps P to Q, namely T_{PQ}.
   (b) As an example, consider points P and Q. Notice that the center of rotation can be any point along the perpendicular bisector of $\overline{PQ}$.

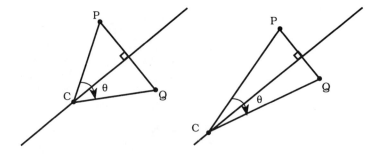

There are infinitely many rotations which transform P to Q. Each has center of rotation C on the perpendicular bisector of $\overline{PQ}$ and an angle of rotation congruent to $\angle PCQ$.

(c) Since the reflection line must be perpendicular to the line joining any point and its image, and it must be equidistant from the point and its image, there is only one reflection line. The reflection line is the perpendicular bisector of $\overline{PQ}$.

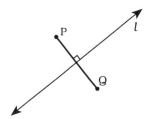

(d) Consider points P and Q. The reflection line and the directed line segment of the translation must be parallel in a glide reflection. Draw any line l through Q. Let this line be the translation line. That is, the translation will be along this line. The reflection line must be parallel to l in such a way that when P is reflected, its image, S, falls on line l. This implies that the reflection line is equidistant from P and the line l. Therefore, the reflection line must go through the midpoint of $\overline{PQ}$. Draw the reflection line through the midpoint of $\overline{PQ}$. Remember that this line must also be parallel to the line l you drew through Q. Now reflect P across this line to some point S. The translation which maps S to Q will be T_{SQ}. The glide reflection consists of a reflection across line l followed by the translation T_{SQ}. Because there are infinitely many lines through point Q, there are infinitely many glide reflections that map P to Q.

Section 16.2

24. (a) To find the center of the size transformation, construct lines through each point and its image. The center of the size transformation will be the point where these lines intersect.

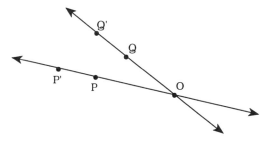

(b) Since OP$'$ > OP and OQ$'$ > OQ, $k > 1$.

(c) Construct lines through each point and its image. The intersection point of the lines is the center of the size transformation.

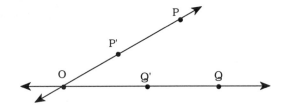

Since OP'< OP and OQ'< OQ, k < 1.

Section 16.2

25. The center of the size transformation will be on $\overleftrightarrow{QQ'}$, and

it will be the point where $\overleftrightarrow{QQ'}$ intersects line l. Since the center of the size transformation, P, is on l, and point R is on l, the transformation of R is also on l. If we connect points R and Q and transform $\overline{RQ}$ using $S_{P,k}$, then the line containing $\overline{R'Q'}$ will be parallel to $\overline{RQ}$ since size transformations preserve parallelism by the theorem which states properties of size transformations. Therefore, the point R' must be the intersection of l and the line through Q' parallel to $\overline{RQ}$.

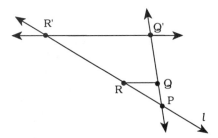

Section 16.2

26. Construct $\overline{PR}$ and $\overline{QR}$. Since size transformations preserve parallelism, R' must be on the line through P' parallel to $\overline{PR}$. Construct this line. Also, R' must be on the line through Q' parallel to $\overline{QR}$. Construct this line. These lines intersect at only one point since they are not parallel, and they are not the same line. Therefore, their point of intersection will be R'.

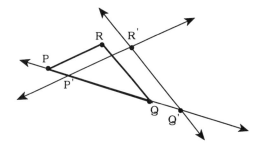

Section 16.2

27. Let R be the point of intersection of line l and $\overline{PP'}$. Let S be the point of intersection of line l and $\overline{QQ'}$.

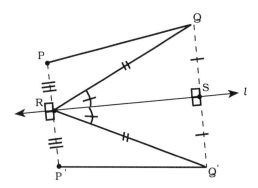

Then $\overline{QS} \cong \overline{Q'S}$ and $\angle QSR \cong \angle Q'SR$ by definition of reflections. We know that $\overline{RS} \cong \overline{RS}$. Therefore, $\triangle QSR \cong \triangle Q'SR$ by SAS. Since corresponding parts of congruent triangles are congruent, we know that $\overline{RQ} \cong \overline{RQ'}$, and $\angle QRS \cong \angle Q'RS$. By the definition of a reflection, we know that $\angle PRS \cong \angle P'RS$, so $m\angle PRS = m\angle P'RS$. Since $m\angle PRS = m\angle PRQ + m\angle QRS$, and $m\angle P'RS = m\angle P'RQ' + m\angle Q'RS$, we have $m\angle PRQ + m\angle QRS = m\angle P'RQ' + m\angle Q'RS$. Also, $m\angle PRQ + m\angle Q'RS = m\angle P'RQ' + m\angle Q'RS$ (since $\angle QRS \cong \angle Q'RS$). We have $\angle PRQ = \angle P'RQ'$ (subtracting $\angle Q'RS$), so $\angle PRQ \cong \angle P'RQ'$. Finally, since $\overline{PR} \cong \overline{P'R}$, by definition of a reflection, $\angle PRQ \cong \angle P'RQ'$, and $\overline{RQ} \cong \overline{RQ'}$ from above, $\triangle PRQ \cong \triangle P'RQ'$ by SAS. Therefore, by corresponding parts, $\overline{P'Q'} \cong \overline{PQ}$, so P'Q' = PQ.

Section 16.3

11. Let M be the midpoint of diagonal $\overline{AC}$. Let H_M be the half turn centered at M.

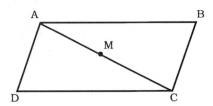

Since H_M is an isometry, distances are preserved, so $H_M(A) = C$ and $H_M(C) = A$. Since ABCD is a parallelogram, $\overline{AB} \| \overline{DC}$. Isometries also maintain parallelism, so $H_M(B)$ is on $\overleftrightarrow{CD}$ since $H_M(\overleftrightarrow{AB}) \| \overleftrightarrow{AB}$ and C is on $H_M(\overleftrightarrow{AB})$.

Also, $H_M(B)$ is on $\overleftrightarrow{AD}$, since $H_M(\overleftrightarrow{BC}) \| \overleftrightarrow{BC}$, and A is on $H_M(\overleftrightarrow{BC})$. Therefore, $H_M(B)$ is the intersection of $\overleftrightarrow{CD}$ and $\overleftrightarrow{AD}$, so $H_M(B) = D$. $H_M(\triangle ABC) = \triangle CDA$, so $\triangle ABC \cong \triangle CDA$.

Section 16.3

12. (a) From question 11, we know $H_M(\triangle ABC) = \triangle CDA$, so $\triangle ABC \cong \triangle CDA$. Therefore, $\angle CDA \cong \angle ABC$, since they are corresponding parts of congruent triangles. We proved previously that consecutive angles of a parallelogram are supplementary. We know then that $m\angle DAB + m\angle ABC = 180°$, and $m\angle BCD + m\angle CDA = 180°$. By substitution, since $\angle CDA \cong \angle ABC$, we know that $m\angle DAB + m\angle CDA = m\angle BCD + m\angle CDA = 180°$. Therefore, $m\angle DAB = 180° - m\angle CDA = m\angle BCD$, so $m\angle DAB = m\angle BCO$, and $\angle DAB \cong \angle BCD$. Opposite angles of a parallelogram are congruent.

    (b) From question 11, $H_M(\overline{AB}) = \overline{DC}$, so $\overline{AB} \cong \overline{DC}$. Similarly $\overline{AD} \cong \overline{BC}$.

Section 16.3

13. Let P be a point on $\overline{AC}$ so that $\overline{BP}$ is the bisector of $\angle ABC$. Let l be the line $\overleftrightarrow{BP}$. Let M_l be the reflection over the line l. Since isometries maintain angle measure, $M_l(A)$ must be on $\overline{BC}$. We know AB = BC, so $M_l(A) = C$.

Thus, $M_l(\triangle ABP) = \triangle CBP$, since $M_l(B) = B$ and $M_l(P) = P$, as points B and P are on the reflection line. Therefore, $\triangle ABP \cong \triangle CBP$. Since corresponding parts are congruent, $\overline{AP} \cong \overline{CP}$, and $\angle BPA \cong \angle BPC$. Notice that these two angles are also supplementary, so they must both be right angles. Therefore, B is on the perpendicular bisector of $\overline{AC}$.

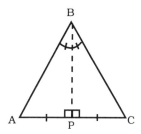

Section 16.3

14. (a) In kite ABCD construct diagonals $\overline{AC}$ and $\overline{DB}$. From question 13, we know that A is on the perpendicular bisector of $\overline{DB}$, and C is on the perpendicular bisector of $\overline{DB}$. Therefore, $\overline{AC} \perp \overline{BD}$.

   (b) The kite will have reflection symmetry if there is a reflection that maps the kite onto itself. We know that $\overline{AC} \perp \overline{BD}$. Because isometries maintain distance and angle measure, $M_{AC}(B) = D$, and $M_{AC}(D) = B$. Since A and C lie on the reflection line, $M_{AC}(A) = A$ and $M_{AC}(C) = C$. Therefore, $M_{AC}(ABCD) = ADCB$, and the kite has reflection symmetry.

Section 16.3

15. $\triangle ABC$ is isosceles. We know that $\overline{AB} \cong \overline{AC}$. $\overleftrightarrow{AP}$ is the bisector of $\angle BAC$. Let l be the line $\overleftrightarrow{AP}$. The image $M_l(B)$ is on $\overleftrightarrow{AC}$, since $\angle BAP \cong \angle CAP$, and reflections maintain angle measure. Since $\overline{AB} \cong \overline{AC}$, and isometries preserve distance, $M_l(B) = C$ and $M_l(C) = B$ so $M_l(\triangle ABP) = \triangle ACP$. Therefore, $\triangle ABP \cong \triangle ACP$. By corresponding parts, we know that $\angle ABP \cong \angle ACP$.

Section 16.3

16. (a) Let A be any point that is x units from r. Since translations are isometries, they preserve distance. Therefore, the distance from A to A' where $A' = M_r(A)$ is $2x$. Similarly, if A' is a distance of y from s, then the distance from A' to A'' where $A'' = M_s(A')$ is $2y$. The distance from A to A'' is $2x + 2y = 2(x + y)$. Because $\overline{AA''} \perp r$, and $r \parallel s$ and $\overline{A'A''} \perp s$ by properties of reflections and parallel lines, we know that A, A', and A'' are collinear. Hence M_r followed by M_s is equal to translation $T_{AA''}$. Since A was an arbitrary point, M_r followed by M_s is $T_{AA''}$.

    (b) In a reflection, the reflection line is the perpendicular bisector of the segment joining a point and its image. If we were to translate the point to its image, then it would move perpendicular to the line. The direction of the translation is perpendicular to both r and s. The distance from A to A'' is $2(x + y)$ [from part (a)]. The distance from A' to s is y. The distance from line r to line s is $x + y$. Therefore the distance for the translation is 2 times the distance between r and s, or $2(x + y)$.

Sectoion 16.3

17. (a) Consider example 16.14. If we want to make cue ball A carom once and hit ball B, we reflect A over the side we want to hit and connect the image of A to point B. The point at which $\overline{A'B}$ intersects the side is the point at which we aim. In this problem, we want ball A to carom off two sides before hitting ball B. We will need two reflections to figure out where to aim. Pick any two consecutive sides off which to carom the ball. For example, choose the bottom and right side. Let l be the line containing the right side, and m be the line containing the bottom side of the pool table. Reflect B over l to point B', where $B' = M_l(B)$. Next reflect B' over m to point B'', where $B'' = M_m(B')$. If we connect A to B'', then we can find point P, where $\overline{AB''}$ intersects the bottom of the table. If we connect P to B', then we can find point Q, where $\overline{PB'}$ intersects the right side of the table. By aiming the ball at point P, the following successful shot will occur.

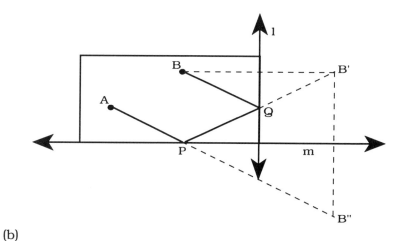

(b)

By the definition of a reflection, we know that $\overline{BS} \cong \overline{B'S}$, and $\angle BSQ \cong \angle B'SQ$. Since $\overline{SQ} \cong \overline{SQ}$, we have $\triangle BSQ \cong \triangle B'SQ$ by SAS. By corresponding parts, we know that $\angle 1 \cong \angle 2$. By vertical angles, $\angle 2 \cong \angle 3$. By substitution, $\angle 1 \cong 3$, so the angle of incidence is equal to the angle of reflection. Similarly, $\triangle B'PT \cong \triangle B''PT$, so $\angle 4 \cong \angle 6$. Therefore, the angle of incident is equal to the angle of reflection for the first carom. If ball A is aimed toward P, then it will bounce off Q and hit ball B.

Section 16.3

18. Consider $S_{A,\frac{b}{a}}$. Since size transformations preserve angle measure, the square ABCD would have image A'B'C'D', which is also a square but with side lengths $a\left(\dfrac{b}{a}\right) = b.$ A'B'C'D' would be congruent to EFGH. By the definition of

congruent shapes, if two shapes, in this case squares, are congruent, then there is an isometry, J, which will map A'B'C'D' to EFGH. Therefore, the combination of $S_{A, \frac{b}{a}}$ and J is the desired similarity transformation.

Section 16.3

19. In order to rotate the cube so that the line segment $\overline{AB}$ is in the same place as the line segment $\overline{DC}$ was and vice versa, we must recall the axes of symmetry of a cube from Chapter 12. Use a model to verify that if the cube is rotated about the axis of symmetry that passes through the midpoints of sides $\overline{PQ}$ and $\overline{RS}$, then the image of $\overline{AB}$ will be $\overline{DC}$, and the image of $\overline{DC}$ will be $\overline{AB}$. Notice that the image of X is still X, and the image of Y is still Y under this rotation. Therefore, X and Y must lie on the axis of rotation. Hence, X and Y are midpoints of the sides $\overline{PQ}$ and $\overline{RS}$, respectively. Since A, B, Y, C, D, and X are all midpoints, the equilateral hexagon has six rotation symmetries. Thus, it must be a regular hexagon. If you rotate hexagon ABYCDX about the center of the cube, there are six rotation symmetries.